Medicinal Fruits of the Middle East

Medicinal plants, including fruits, contain bioactive components that present health benefits for humans. There are several medicinal fruits, including jackfruit, pomegranate, date palm, papaya, and lemons, grown commercially in countries throughout the Middle East. *Medicinal Fruits of the Middle East* explores this subject by featuring information on bioactive properties, extraction techniques, natural colorants, nutraceutical preparations, and traditional food preparations of these unique medicinal fruits.

A volume in the Nutraceuticals: Basic Research and Clinical Applications series, *Medicinal Fruits of the Middle East* presents a wealth of knowledge from chapter authors versed in the local traditions and expert knowledge of these fruits. It is a key text for food scientists, engineers, and nutritionists, as well as researchers, academicians, and those who work in the functional foods and nutraceuticals industry.

Nutraceuticals: Basic Research and Clinical Applications

Series Editor: Yashwant Pathak, PhD

Nutraceuticals for Aging and Anti-Aging: Basic Understanding and Clinical Evidence
edited by Jayant Lokhande and Yashwant Pathak

Marine-Based Bioactive Compounds: Applications in Nutraceuticals
edited by Stephen T. Grabacki, Yashwant Pathak, and Nilesh H. Joshi

Applications of Functional Foods and Nutraceuticals for Chronic Diseases: Volume I
edited by Syam Mohan, Shima Abdollahi, and Yashwant Pathak

Flavonoids and Anti-Aging: The Role of Transcription Factor Nuclear Erythroid 2-Related Factor 2
edited by Karam F.A. Soliman and Yashwant Pathak

Molecular Mechanisms of Action of Functional Foods and Nutraceuticals for Chronic Diseases: Volume II
edited by Shima Abdollahi, Syam Mohan, and Yashwant Pathak

Plant Food Phytochemicals and Bioactive Compounds in Nutrition and Health
edited by John O. Onuh and Yashwant V. Pathak

Anxiety, Gut Microbiome, and Nutraceuticals: Recent Trends and Clinical Evidence
edited by Yashwant V. Pathak, Sarvadaman Pathak, and Con Stough

Wealth Out of Food Processing Waste: Ingredient Recovery and Valorization
edited by Meenakshisundaram Selvamuthukumaran

Nutraceuticals Inspiring the Contemporary Therapy for Lifestyle Diseases
edited by Mala Trivedi, Sachidanand Singh, Parul Johri, and Pedro Lopez-Sanchez

Medicinal Fruits of the Middle East
edited by Meenakshisundaram Selvamuthukumaran

For more information about this series, please visit:
routledge.com/Nutraceuticals/book-series/CRCNUTBASRES

Medicinal Fruits of the Middle East

Edited by
Meenakshisundaram Selvamuthukumaran

CRC Press
Taylor & Francis Group
Boca Raton London New York

CRC Press is an imprint of the
Taylor & Francis Group, an **Informa** business

Designed cover image: Shutterstock

First edition published 2026
by CRC Press
2385 NW Executive Center Drive, Suite 320, Boca Raton FL 33431

and by CRC Press
4 Park Square, Milton Park, Abingdon, Oxon, OX14 4RN

CRC Press is an imprint of Taylor & Francis Group, LLC

ISBN: 9781032589428 (hbk)
ISBN: 9781032590288 (pbk)
ISBN: 9781003452614 (ebk)

DOI: 10.1201/9781003452614

Typeset in Palatino
by codeMantra

Dedication

I profoundly thank

God

My Family

My Friends

and

Everyone

Who have inspired, supported, and wholeheartedly encouraged me to complete this book

Meenakshisundaram Selvamuthukumaran

Contents

About the editor ix
Contributors x
Introduction xii

Chapter 1 Exploration of medicinal fruits of Saudi Arabia 1
Meenakshisundaram Selvamuthukumaran

Chapter 2 Processing of functional foods from medicinal fruits of UAE 7
Meenakshisundaram Selvamuthukumaran

Chapter 3 Extraction of natural colourants from medicinal fruits of Iran 14
Riya Goyal and Vikas Nanda

Chapter 4 Extraction of antioxidants from medicinal fruits of Iraq 56
Mehvish Habib, Kulsum Jan, Eram S. Rao, and Khalid Bashir

Chapter 5 Therapeutic potential of medicinal fruits of Lebanon 116
Meenakshisundaram Selvamuthukumaran

Chapter 6 Value addition of medicinal fruits of Muscat 122
Divya Chauhan and Sachin Kumar

Chapter 7 Development of nutraceuticals from medicinal fruits of Bahrain 132
K. Yogeshwari, Shivani Chauhan, Gurdeep Rattu, Divya Chauhan, and Sachin Kumar

Chapter 8 Bioactive properties of processed products of medicinal fruits from Kuwait 146
Rohini Dhenge, Arjun Mohanakumar, Massimiliano Rinaldi, and Tommaso Ganino

Chapter 9 Bioavailability of therapeutic constituents of medicinal fruits of Jordan 176
Meenakshisundaram Selvamuthukumaran

Chapter 10 Traditional/ethnic food preparations from medicinal fruits of Egypt 180
Arjun Mohanakumar, Rohini Dhenge, and Massimiliano Rinaldi

Index 193

About the editor

Dr. Meenakshisundaram Selvamuthukumaran is currently working as a Professor and Dean of the School of Agriculture, Joy University, Vadakangulam, Tamil Nadu. He was a Visiting Professor at the School of Food Science and Postharvest Technology, Institute of Technology, Haramaya University, Dire Dawa, Ethiopia, and also at Hamelmalo Agricultural College, Eritrea. Dr. Selvamuthukumaran earned his Ph.D. in Food Science from the Defence Food Research Laboratory, affiliated with the University of Mysore, India. His core area of research is the processing of underutilized fruits for the development of antioxidant-rich functional food products. He has transferred several technologies to Indian firms as an outcome of his research work. He has received several awards and citations for his research work. Dr. Selvamuthukumaran has published international papers and book chapters in the area of antioxidants and functional foods. He has guided several national and international postgraduate students in the area of food science and technology.

Contributors

Khalid Bashir
Department of Food Technology
Jamia Hamdard
New Delhi, Delhi, India

Divya Chauhan
Department of Food Technology
Eternal University
Baru Sahib, Himachal Pradesh, India

Shivani Chauhan
Department of Microbiology
University of Delhi South Campus
New Delhi, Delhi, India

Rohini Dhenge
Department of Food and Drug
University of Parma
Parma, Italy

Tommaso Ganino
Department of Food and Drug
University of Parma
Parma, Italy

Riya Goyal
Department of Food Engineering and Technology
Sant Longowal Institute of Engineering and Technology
Punjab, India

Mehvish Habib
Department of Food Technology
Jamia Hamdard
New Delhi, Delhi, India

Kulsum Jan
Department of Food Technology
Jamia Hamdard
New Delhi, Delhi, India

Sachin Kumar
Department of Food Process Technology
Banda University of Agriculture and Technology
Uttar Pradesh, India

Arjun Mohanakumar
Department of Food and Drug
University of Parma
Parma, Italy

Vikas Nanda
Department of Food Engineering and Technology
Sant Longowal Institute of Engineering and Technology
Punjab, India

Eram S. Rao
Department of Food Technology
Bhaskaracharya College of Applied Sciences
University of Delhi
New Delhi, Delhi, India

Gurdeep Rattu
National Horticulture Research and Development Foundation (NHRDF)
Maharashtra, India

Massimiliano Rinaldi
Department of Food and Drug
University of Parma
Parma, Italy

K. Yogeshwari
Department of Biology
BASE PU College
Bengaluru, India

Introduction

Chapter 1 deals with the exploration of various medicinal fruits found in Saudi Arabia, such as Bamber *(Cordia myxa)*, Jakiya (Jackfruit), Jumaiza (Sycamore), Tarthuth *(Cynomorium coccineum)*, Atar/Al-Jarawa *(Glossonema varians)*, and Al-Qurayshman/Al-Mughath *(Glossostemon bruguieri)*. Also, the cultivation of these medicinal fruits and their therapeutic role will be discussed in detail. Chapter 2 deals with the processing of various functional food products from medicinal fruits of the UAE, such as Desert squash, apricot, and papaya. Chapter 3 describes the extraction of natural colorants such as anthocyanins and carotenoids from medicinal fruits of Iran, such as fig, date palm, and pomegranate, using different processing techniques. Chapter 4 narrates the extraction of antioxidants from medicinal fruits of Iraq, such as *Foeniculum vulgare* (Habit Helwa), *Carum carvi* (Carawaya), *Pimpinella anisum* (Yansoon), and *Rhus coriaria* (Summak). Thermal and non-thermal extraction methods for antioxidants are discussed. Chapter 5 deals with the therapeutic potential of medicinal fruits of Lebanon, such as kiwi, jujube, pomegranate, and fig. Chapter 6 deals with adding value to medicinal fruits that are grown in Oman, such as dates, fig, pomegranate, and grapes. Chapter 7 describes the development of nutraceuticals from medicinal fruits of Bahrain, such as lemon, *Ficus palmata, Ziziphus spina-christi, Sageretia thea, Grewia tenax,* and *Grewia erythraea*. Chapter 8 narrates the bioactive properties of processed products of medicinal fruits from Kuwait, such as lucky bamboo, puncture vine, cheeseweed mallow, sweet basil, Indian laurel, and London rocket. Chapter 9 explains the bioavailability of therapeutic constituents of medicinal fruits of Jordan, such as pomegranate and *Ecballium* elaterium. Chapter 10 narrates the traditional/ethnic food preparations from medicinal fruits of Egypt, such as *Terminalia bellerica, Terminalia chebula,* and *Terminalia horrida.*

chapter one

Exploration of medicinal fruits of Saudi Arabia

Meenakshisundaram Selvamuthukumaran

1.1 Introduction

The Saudi Arabian Kingdom has excellent, unexplored rare fruits, which possess significant medicinal properties (Table 1.1). The various medicinal fruits that exist in the Arabian Kingdom, along with their therapeutic effects, are described below.

1.1.1 Cordia myxa

Cordia myxa, commonly known as Assyrian plums, belongs to the family Borage, growing abundantly in continents such as Asia, Africa, and America. It consists of broad leaves, and it is a deciduous tree that possesses rich medicinal values (Schmelzer and Gurib-Fakim, 2008; Warfa, 1988). The tree starts producing fruit in August, which has a pale color, and when the fruit reaches the fully ripened stage, the pulp becomes translucent with a sweeter taste (Schmelzer and Gurib-Fakim, 2008; Warfa, 1988). This fruit is utilized for medicinal purposes in Middle East countries such as Saudi Arabia, and it is also highly used in other countries such as China and India because of its demulcent properties. The fruit's unique medicinal properties have been exploited to treat coughs, sore throats, respiratory infections, etc.

The fruit is used as a diuretic, and the pulp obtained from this fruit is utilized effectively for treating rheumatic pain. It is also used as an anthelminthic (Alami and Macksad, 1974; Rechinger, 1997; Schmelzer and Gurib-Fakim, 2008). Different authors have reported that the fruit possesses significant bio-active properties, such as immunomodulatory, antiparasitic, antimicrobial, analgesic, cardiovascular, respiratory, insecticidal, gastrointestinal, and protective effects (Rashed et al., 2015; Pandey et al., 2014; Abdallah et al., 2011). Besides these medicinal properties, the fruit is

DOI: 10.1201/9781003452614-1

Table 1.1 Medicinal properties of Saudi Arabian fruits

Common name of the fruit	Scientific name	Medicinal value	References
Assyrian plums	*Cordia myxa*	Immunomodulatory, antiparasitic, antimicrobial, analgesic, cardiovascular, respiratory, insecticidal, gastrointestinal, and protective effects	Rashed et al. (2015), Pandey et al. (2014), and Abdallah et al. (2011)
Dombeya arabica Baker	*Glossostemon bruguieri*	Demulcent agent	Gamel et al. (2010)
Attar	*Glossonema varians*	Flavonoids in fruits contribute to nutraceutical value	Rizk et al. (1983)
Red thumb	*Cynomorium coccineum*	Presence of phenolic compounds contributes to antioxidant activity	Zucca et al. (2013)
Sycamore fig	*Ficus sycomorus*	Antiviral, antioxidant, antimicrobial, cardioprotective, and hepatoprotective properties	Salvador et al. (2012), Laszczyk (2009), Yan et al. (2014), and Dzubak et al. (2006)

also a good source of proteins, carbohydrates, fibers, etc. Some researchers recommend using this fruit as a nutritive supplement, especially for the livelihood of poor natural communities (Al-Snafi, 2017).

1.1.1.1 Bio-active compounds identified from Cordia myxa *fruit*

Cordia myxa fruit possesses significant bio-active components that can be quantified, which are described below and also given in Figure 1.1.

1.1.1.1.1 Stigmasterol It is one of the important components quantified from the fruit *Cordia myxa*. It plays a major role in inhibiting different types of cancers, such as breast, prostate, ovarian, and colon. Other medicinal properties, including hypoglycemic, antithyroid protective effects, and hyperlipidemic properties, have been well described by various authors (Woyengo et al., 2009; Panda et al., 2009).

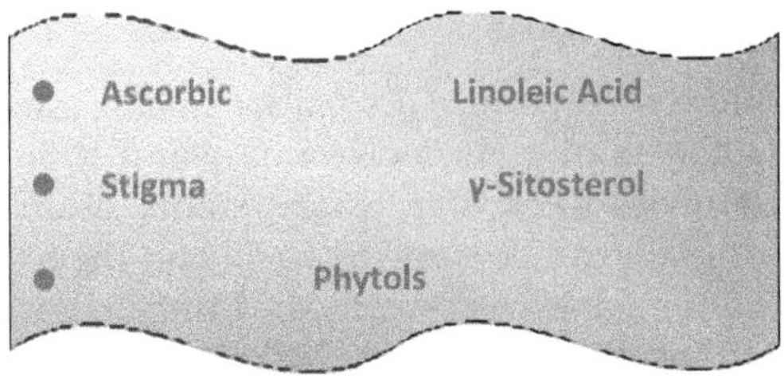

Figure 1.1 Bioactive components quantified from *Cordia myxa* fruit.

1.1.1.1.2 Phytol The presence of phytol in the fruit can regulate blood glucose levels, thereby efficiently restoring the metabolic functions of Type II diabetes (Elmazar et al., 2013; National Center for Biotechnology Information, 2017). It showed good cytotoxic and anti-proliferative effects when tested on human cells.

1.1.1.1.3 Linoleic acid This acid helps in effectively controlling weight, inhibiting blood clotting, and thereby significantly possessing both anti-inflammatory and pro-inflammatory effects (Attar-Bashi et al., 2004).

1.1.1.1.4 Ascorbic acid The ascorbic acid present in the fruit can minimize the intensity of cancer, increase immunity, and regulate blood pressure. Its presence can stimulate enzymes, activate hormones, and detoxify histamine (Shwaish and Al-Imarah, 2017).

1.1.1.1.5 γ-Sitosterol The presence of γ-Sitosterol in the fruit can reduce the level of cholesterol absorption from various dietary sources (Heinemann et al., 1991).

1.1.2 Glossostemon bruguieri

Dombeya arabica Baker is scientifically known as *Glossostemon bruguieri* or *Dombeya arabica*, which is a flowering plant belonging to the family Malvaceae, widely growing in Middle East countries including Saudi Arabia, Iraq, Iran, Morocco, Egypt, Yemen, and Turkey (Meselhy and Meselhy, 2003; Gamel et al., 2010).

It's a shrub with longer roots that are darker in color, and it can extend up to 70–100 cm in length, with a breadth of 5–8 cm (Purseglove, 1974). Seeds of this shrub are found to contain 19.5% protein, 1.8% arabinose, and 14.6% glucuronic acid. Roots obtained from this shrub are used as medicine, i.e., as a demulcent agent for the relief of especially gout pain (Gamel et al., 2010).

1.1.3 Glossonema varians

It is commonly known as Attar, *Glossonema edule*, jarawa, and yarawa, which belongs to the family Asclepiadaceae, and it is found growing in Middle East countries like Qatar, Saudi Arabia, etc. The tender fruits obtained from this plant can be eaten as such, and they have good nutraceutical value. The tender leaves, seeds, and unripe fruits are also consumed and utilized locally (Singh et al., 2017). The fruits contain essential amino acids, fatty acids, and flavonoids (Rizk et al., 1983).

1.1.4 Cynomorium coccineum

It's a parasitic perennial flowering plant that belongs to the family Cynomoriaceae. These plants are abundantly found in the desert areas of Middle East countries like Saudi Arabia, Jordan, etc. It is commonly called the Desert Thumb or the Red Thumb. It contains various active ingredients like lignans, triterpenoid saponins, anthocyanin glycosides, cyanidin-3-O-glucoside, and gallic acid (Zucca et al., 2013).

1.1.5 Ficus sycomorus

It is known as sycamore fig or fig mulberry, as the leaves look like mulberry, and it belongs to the Moraceae family. It's a fig species widely grown in Saudi Arabia, Egypt, and Lebanon. It has been cultivated since ancient times. The fruit is edible, appearing in red or yellow color with a diameter of 2–3 cm. Fruiting occurs throughout the year, with the peak period starting from July until December.

Several bio-active components can be extracted from this fig species, which are derivatives of flavonoids, phenols, and terpenoids (Yan et al., 2014; Dzubak et al., 2006; Chiang et al., 2005; Chiang and Kuo, 2001). These components can exhibit various bio-active properties, such as antiviral, antioxidant, antimicrobial, and cardio- and hepatoprotective properties (Salvador et al., 2012; Laszczyk, 2009; Yan et al., 2014; Dzubak et al., 2006).

1.2 *Conclusions*

There is a wide variety of rare fruits that can be explored in Saudi Arabia, and these fruits have potent bioactive properties. Their ancient historical findings reveal that they can be successfully used as medicine for treating several disorders.

References

Abdallah, I.A., Khattab, H.H., and Heeba, G.H. (2011). Gastroprotective effect of *Cordia myxa* L. fruit extract against indomethacin-induced gastric ulceration in rats. *Life Science Journal*, 8(3):433–445.

Alami, R. and Macksad, A. (1974). *Medicinal Plants in Kuwait* (Volume 6). Kuwait: Al-Assriya Press Volume, pp. 50–137.

Al-Snafi, A.E. (2017). The Pharmacological and therapeutic importance of *Cordia myxa*. *IOSR Journal of Pharmacy*, 6(6):47–57.

Attar-Bashi, N.M., Li, D., and Sinclair, A.J. (2004). Alpha-linolenic acid and the risk of prostate cancer. *Lipids*, 39(9):929–32.

Chiang, Y.M., Chang, J.Y., Kuo, C.C., Chang, C.Y., and Kuo, Y.H. (2005). Cytotoxic triterpenes from the aerial roots of *Ficus microcarpa*. *Phytochemistry*, 66(4): 495–501.

Chiang, Y.M., and Kuo, Y.H. (2001). New peroxy triterpenes from the aerial roots of *Ficus microcarpa*. *Natural Product Reports*, 64(4):436–439.

Dzubak, P., Hajduch, M., Vydra, D., Hustova, A., Kvasnica, M., and Biedermann, D. (2006). Pharmacological activities of natural triterpenoids and their therapeutic implications. *Natural Product Reports*, 23(3):394–411.

Elmazar, M.M., El-Abhar, H.S., Schaalan, M.F., and Farag, N.A. (2013). Phytol/phytanic acid and insulin resistance: potential role of phytanic acid proven by docking simulation and modulation of biochemical alterations. *PLoS One*, 8(1):e45638.

Gamel, T.H., Abd El-Razek, A.M., and Damir, A.A. (2010). Dried peeled roots of *Glossostemon bruguieri* (Moghat) as a potential functional food. *Journal of Food Processing and Preservation*, 34(1):55–67.

Heinemann, T., Kullak-Ublick, G.A., Pietruck, B., and von Bergmann, K (1991). Mechanisms of action of plant sterols on inhibition of cholesterol absorption. Comparison of sitosterol and sitostanol. *European Journal of Clinical Pharmacology*, 40(Suppl 1):S59–S63.

Laszczyk, M.N. (2009). Pentacyclic triterpenes of the lupane, oleanane and ursane group as tools in cancer therapy. *Planta Medica*, 75(15):1549–1560.

Meselhy, R.M. (2003). Constituents from Moghat, the roots of *Glossostemon bruguieri* (Desf.). *Molecules*, 8(8):614–621.

National Center for Biotechnology Information. (2017). PubChem Compound Database (CID=5280435). Available online: https://pubchem.ncbi.nlm.nih.gov/compound/5280435 (accessed 5 March 2017).

Panda, S., Jafri, M., Kar, A., and Meheta, B.K. (2009). Thyroid inhibitory, antiperoxidative and hypoglycemic effects of stigmasterol isolated from Butea monosperma. *Fitoterapia*, 80(2):123–126.

Pandey, B., Deshpand, B., Singh, S., and Chandrakar, V. (2014). Estimation of elemental contents of *Cordia myxa* and its antimicrobial activity against various pathogenic microorganisms. *Indian Journal of Scientific Research*, 4(1):39–44.

Purseglove, J.W. (1974). *Tropical Crops, Dicotyledons*, 3rd ed.; Wiley: New York.

Rashed, K., Luo, M.T., Zhang, L.T., and Zheng, Y.T. (2015) Evaluation of anti-HIV-activity of *Cordia myxa* L., and phytochemical profile. *Banat's Journal of Biotechnology*, 5(10):75–82.

Rechinger, K.H. (1997). Boraginaceae. In: Riedl, H., (Ed.) *Flora Iranica* (Volume 48). Graz: Akademische Durck-u Verlangsantalt, pp. 6–10.

Rizk, A.M., Hammouda, F.M., and Hussein, L. (1983). Constituents of plants growing in Qatar III-nutritive constituents of Garawah (Glossonema edible). *Plant Foods Human Nutrition*, 33:71–76.

Salvador, J.A.R., Moreira, V.M., Gonçalves, B.M.F., Lealab, A.S., and Jing, Y. (2012). Ursane-type pentacyclic triterpenoids as useful platforms to discover anti-cancer drugs. *Natural Product Reports*, 29(12):1463–1479.

Schmelzer, G. H., Gurib-Fakim, A., Arroo, R., Bosch, C. H., de Ruijter, A., Simmonds, M. S. J., Lemmens, R. H. M. J., & Oyen, L. P. A. (2008). *Plant Resources of Tropical Africa 11(1): Medicinal plants 1*. (Plant Resources of Tropical Africa; No. 11(1)). PROTA. Wageningen: Earthprint Limited, p. 791. https://edepot.wur.nl/417238

Shwaish, T., and Al-Imarah, F.J.M. (2017). Chemical composition of Cordia Myxa fruit: phytochemical screening and identification of some bioactive compounds. *International Journal of Advanced Research*, 5(9), 1255–1260.

Singh, J.P., Kumar, S., and Kulloli, R.N. (2017). Glossonema variens: a threatened food cum medicinal plant of western Rajasthan. *Indian Journal of Arid Horticulture*, 12(1–2):16–18.

Warfa, A.M. (1988). Cordia (Boraginaceae). In: *NE Tropical Africa and Tropical Arabia*. Uppsla: Uppsla Univ., p. 174.

Woyengo, T.A., Ramprasath, V.R., and Jones, P.H. (2009). Anticancer effects of phytosterols. *European Journal of Clinical Nutrition*, 63:813–820.

Yan, X.J., Gong, L.H., Zheng, F.Y., Cheng, K.J., Chen, Z.S., and Shi, Z. (2014). Triterpenoids as reversal agents for anticancer drug resistance treatment. *Drug Discovery Today*, 19(4):482–488.

Zucca, P., Rosa, A., Tuberoso, C.I.G., Piras, A., Rinaldi, A.C., Sanjust, E., Dessì, M.A., and Rescigno, A. (2013). Evaluation of antioxidant potential of "Maltese mushroom" (*Cynomorium coccineum*) by means of multiple chemical and biological assays. *Nutrients*, 5(1):149–161.

chapter two

Processing of functional foods from medicinal fruits of UAE

Meenakshisundaram Selvamuthukumaran

2.1 Introduction

The United Arab Emirates grows several fruit crops, which have unique medicinal properties. The various health benefits associated with the cultivation, growing, and value addition of these medicinal fruit crops are described below:

2.2 Papaya

Papaya fruit, scientifically known as *Carica papaya*, is growing widely in the United Arab Emirates nowadays. A lot of phytoconstituents are present in the pulp, seed, and skin of the papaya fruit, including polyphenols and carotenoids (Rivera-Pastrana et al., 2010). The yellow-flush varieties of papaya contain functional pigments like lutein, beta-carotene, and carotenoids, while the red-flush varieties contain lycopene (Shen et al., 2019). The papaya fruit contains dietary fibre of around 2%, with significant proportions of vitamin C (around 45 mg), which can significantly contribute to 75% of the daily value. The presence of functional pigments in the fruit has significantly contributed to several bioactive properties, including antioxidant activity, anti-cancer activity, anti-inflammatory activity, and improved heart health, ascribed to the presence of natural fruit fibres.

2.3 Development of functional food products from medicinal fruits

2.3.1 Processing of papaya

2.3.1.1 Papaya jam

Good-quality ripened fruit is utilised for preparing papaya fruit jam. The ripened fruit is selected and peeled, the seeds are removed, and it is crushed in a pulper to obtain pulp. The thick pulp obtained from the papaya fruit

DOI: 10.1201/9781003452614-2

serves as a raw material for developing papaya jam (Figure 2.1). The other ingredients used for developing papaya jam are pectin, sugar, citric acid and papaya pulp (Table 2.1). The desired amount of fruit pulp is transferred to a steam-jacketed kettle, where the pulp is boiled to extract natural pectin, followed by the incorporation of the desired amount of sugar. The mixture is then boiled to a thick consistency, so that the Total Soluble Solids (TSS) of the final product reaches 68.5° Brix. The product is finally packed in sterilised glass jars up to the brim, sealed airtight, and stored until it is marketed.

Selection of good quality of ripened papaya fruit

↓

Peeling and removal of seed

↓

Pulping

↓

Boiling

↓

Addition of sugar, citric acid and Pectin

↓

Final boiling to achieve desired TSS

↓

Packing and marketing

Figure 2.1 Flow sheet for preparation of papaya jam.

Table 2.1 Ingredients required for preparing papaya jam

Name of the ingredients	Ingredients quantity (kg)
Sugar	1
Papaya pulp	1
Pectin	0.005
Citric acid	0.005

2.3.1.2 *Papaya fruit juice concentrates*

Fruit juice concentrate is nothing but the process of evaporating the moisture content from the freshly obtained papaya pulp by five to six times, so that the product's stability can be enhanced to a greater extent. The papaya fruit concentrate forms the basis for developing several papaya-based fruit products. The flow sheet for preparing papaya fruit concentrate is given in Figure 2.2.

Papaya fruit

↓

Peeling and slicing

↓

Removal of seeds

↓

Pulping

↓

Concentration of pulp in evaporator until 60-70° Brix is achieved

↓

Packing in sterilized containers

↓

Storing

↓

Marketing

Figure 2.2 Flow sheet for developing papaya fruit juice concentrates.

2.4 Apricot

Apricot is a fruit-bearing small tree, which grows up to a height of 10–12 m. It belongs to the family *Prunus armeniaca*. The apricot contains phytochemicals like provitamin A, polyphenols, beta-carotene, chlorogenic acid, and catechins (Campbell et al., 2013). It also contains high concentrations of potassium and vitamin C.

2.4.1 Processed products from apricot

2.4.1.1 Apricot squash

Apricot squash can be developed from apricot pulp, incorporating other desired ingredients like citric acid and sugar. It's a concentrated product, and it needs reconstitution at a 1:3 ratio before serving. The ingredients required for preparing apricot squash are provided in Table 2.2, and the detailed flow sheet for preparing apricot squash is given in Figure 2.3.

Table 2.2 Ingredients required for developing apricot squash

Name of the ingredients	Quantity (kg)
Apricot pulp	0.300
Sugar	0.450
Citric acid	0.009

Water is needed @ 0.3 L for preparing sugar syrup.

Sorting of apricot fruit

↓

Peeling, slicing and Destoning

↓

Pulping

↓

Preparation of acidified sugar syrup

↓

Blending of acidified sugar syrup with apricot pulp

↓

Addition of preservatives

↓

Packing hot in sterilized glass jars

↓

Sealing

↓

Storing and marketing

Figure 2.3 Flow sheet for preparing papaya fruit squash.

2.4.1.2 *Drying of apricot*

Apricot can be dried efficiently to increase its stability. Drying can be carried out by either sun drying or mechanical drying processes. The sun drying process has several disadvantages because it requires a much longer drying period. On the other hand, the mechanical drying process has several advantages; the product can be dried very quickly, within 20 hours, thereby maintaining the quality of the product to a greater extent. The dried apricot fruit contains higher concentrations of nutrients like vitamin A (180 μg), potassium (1,160 mg), dietary fibre (7 g), and protein (4 g/100 g). The flow sheet for preparing dried apricot fruit by the mechanical dehydration process is described in Figure 2.4.

Selection of good quality of apricot

↓

Cutting into half or using as a whole for drying the fruits

↓

Sulphiting (Dipping the fruit in KMS solution @ 1-2%)

↓

Drying them in tray dryer at a temperature of 50 °C for a period of 20-24 hours

↓

Packing and storing

Figure 2.4 Flow sheet for preparing dried apricot fruit by mechanical dehydration process.

2.5 Citrullus colocynthis

It is commonly known as the bitter apple fruit, and it's a perennial plant with vine-like stems that can spread on the ground and even climb up from there. The plant produces yellow flowers at the leaf axils. These plants are also widely found in the African and Asian continents (Pravin et al., 2013; Li et al., 2022). The United Arab Emirates also has such a unique fruit, which can contribute significant medicinal value. The seed contains fibre (1.9%), carbohydrates (10.9%), protein (25%), and fat (57%) (Ogundele et al., 2012)

2.5.1 *Bioactive properties of* C. colocynthis

Hameed et al. (2020) and Kumar et al. (2008) reported that *Citrullus colocynthis* can be used to cure gastrointestinal, skin, and bacterial infections, cancer, diabetes, and oedema-related problems. Hussain et al. (2014) demonstrated that the dried powder obtained from the pulp derived from this fruit can be used as one of the remedies for indigestion and gastroenteritis-related problems. The extract or decoction obtained from the fruit can be used as an analgesic (Heydari et al., 2015) and the whole plant has potent possibilities to cure liver-related problems and treat intestinal paralysis (Rahimi et al., 2012).

2.5.2 *Limitations of* Citrullus colocynthis *for its value additions*

The fruit has good medicinal properties, but there are many disadvantages that may discourage the use of *Citrullus colocynthis* for its value addition. The main problem associated with the processing of this plant is that it contains some constituents, like cucurbitacin, that can irritate the mucous membranes. The Food and Drug Administration has already banned the use of *Citrullus colocynthis* since 1991, as the whole plant can cause kidney damage, bloody diarrhoea, bloody urine, convulsions, paralysis, gut and stomach irritations, etc. Several research studies show that powder obtained from *Citrullus colocynthis* fruit, even at a low dose concentration of less than 2 g, can cause death. Consumption of powder at a low dose concentration of less than 300 mg can also cause diarrhoea in consumers. Oral administration of *Citrullus colocynthis* fruit at 0.25 g/kg/day resulted in diarrhoea, dehydration, and recumbency before death (Adam et al., 2000).

Therefore, it is highly advisable to take utmost care before consuming either the raw fruit or its processed products. Without prior consultation with a healthcare expert, no one should consider consuming such fruits or their products now or in the near future.

2.6 Conclusions

Various functional processed fruit products like squash, jam, concentrates, and dried products can be successfully prepared from apricots and papaya. However, the rare fruit, like bitter apple, cannot be utilised for further processing because of the presence of toxic constituents in the fruit. One needs to consult a healthcare expert before consuming any natural products obtained from bitter apple.

References

Adam, S. E. I., Al-Farhan, A. H., and Al-Yahya, A. (2000). Effect of combined Citrullus colocynthis and Rhazya stricta use in Najdi sheep. *The American Journal of Chinese Medicine*, 28 (3–4), 385–390.

Campbell, O. E., Merwin, I. A., and Padilla-Zakour, O. I. (2013). Characterization and the effect of maturity at harvest on the phenolic and carotenoid content of Northeast USA Apricot (Prunus armeniaca) varieties. *Journal of Agricultural and Food Chemistry*, 61(51), 12700–12710.

Hameed, B., Ali, Q., Hafeez, M. M., and Malik, A. (2020). Antibacterial and antifungal activity of fruit, seed and root extracts of Citrullus colocynthis plant. *Biological and Clinical Sciences Research Journal*, 2020(1). https://doi.org/10.54112/bcsrj.v2020i1.33

Heydari, M., Homayouni, K., Hashempur, M. H., and Shams, M. (2015). Topical Citrullus colocynthis (bitter apple) extract oil in painful diabetic neuropathy: A double-blind randomized placebo-controlled clinical trial. *Journal of Diabetes*, 8, 246–252.

Hussain, A. I., Rathore, H. A., Sattar, M. Z., Chatha, S. A., Sarker, S. D., and Gilani, A. H. (2014). Citrullus colocynthis (L.) Schrad (bitter apple fruit): A review of its phytochemistry, pharmacology, traditional uses and nutritional potential. *Journal of Ethnopharmacology*, 155(1), 54–66.

Kumar, S., Kumar, D., Manjusha, K., Saroha, K., Singh, N., and Vashishta, B. (2008). Antioxidant and free radical scavenging potential of Citrullus colocynthis (L.) Schrad. Methanolic fruit extract. *Acta Pharmaceutica*, 58(2), 215–220.

Li, Q.-Y., Munawar, M., Saeed, M., Shen, J.-Q., Khan, M. S., Noreen, S., Alagawany, M., Naveed, M., Madni, A., and Li, C.-X. (2022). Citrullus colocynthis (L.) Schrad (bitter apple fruit): Promising traditional uses, pharmacological effects, aspects, and potential applications. *Frontiers in Pharmacology*, 12, 791049.

Ogundele, J. O., Oshodi, A. A., and Amoo, I. A. (2012). Comparative study of amino acid and proximate composition of Citurllus colocynthis and Citrullus vulgaris seeds. *Pakistan Journal of Nutrition*, 11(3), 247.

Pravin, B., Tushar, D., Vijay, P., and Kishanchnad, K. (2013). Review on Citrullus colocynthis. *International Journal of Research in Pharmacy and Chemistry*, 3(1), 46–53. https://doi.org/10.1089/acm.2011.0297

Rahimi, R., Amin, G., and Ardekani, M. R. (2012). A review on Citrullus colocynthis Schrad: from traditional Iranian medicine to modern phytotherapy. *The Journal of Alternative and Complementary Medicine*, 18(6), 551–554.

Rivera-Pastrana, D., Yahia, E., and González-Aguilar, G. (2010). Phenolic and carotenoid profiles of papaya fruit (Carica papaya L.) and their contents under low-temperature storage. *Journal of the Science of Food and Agriculture*, 90(14), 2358–2365.

Shen, Y. H., Yang, F. Y., Lu, B. G., Zhao, W. W., Jiang, T., Feng, L., Chen, X. J., and Ming, R. (2019). Exploring the differential mechanisms of carotenoid biosynthesis in the yellow peel and red flesh of papaya. *BMC Genomics*, 20(1), 49.

chapter three

Extraction of natural colourants from medicinal fruits of Iran

Riya Goyal and Vikas Nanda

3.1 Introduction

Iran is home to a wide variety of therapeutic plants and fruits, as well as a rich history of traditional medicine and biodiversity (Dolatkhahi, Dolatkhahi, & Nejad, 2014). In addition to being essential to the region's medical and cultural traditions, these natural resources have enormous potential for the creation of natural colourants, which are becoming more and more in demand across a range of industries because of their eco-friendliness, safety, and biocompatibility (Aseel, Behiry, & Abdelkhalek, 2023).

For ages, natural colourants sourced from plants have been utilised in food, cosmetics, and textile manufacturing (Hosseinnezhad & Safapour, 2024). However, serious health and environmental issues were raised by the synthetic dyes that were common in the 19th century, which sparked a renewed interest in natural alternatives. Medicinal fruits are a unique source of natural colourants because of their high pigment concentration and therapeutic qualities. Iran's varied geography and special climate allow for the growth of a vast range of therapeutic fruits, each with a distinctive phytochemical content (Hosseini et al., 2022).

Medicinal fruits of significance from Iran that are abundant in natural colourants are black mulberry (*Morus nigra*), pomegranate (*Punica granatum*), barberry (*Berberis vulgaris*), and saffron (*Crocus sativus*). For example, pomegranates are well known for their intensely reddish pigments, called anthocyanins, which are powerful antioxidants (Venkitasamy et al., 2019). Berberine and other alkaloids are abundant in barberry fruits, which account for their rich red colour and therapeutic properties (Hashmi et al., 2024). The stigmas of the *Crocus sativus* flower are used to make saffron, which is considered to be the most expensive spice in the world. It is highly valued for its golden-yellow colour, which is caused by

DOI: 10.1201/9781003452614-3

the carotenoid crocin (Kothari et al., 2021). The rich purple-black colour of black mulberries is another excellent source of anthocyanins (Ahlawat et al., 2016).

Several steps are involved in extracting natural colourants from these fruits, including harvesting, drying, grinding, and solvent extraction. Each step is crucial in determining the yield and purity of the colourants. The pigments are dissolved in a variety of solvents, including water, ethanol, and methanol. Advanced techniques such as supercritical fluid extraction and ultrasound-assisted extraction are also being explored to improve the efficiency and sustainability of the process (Chemat et al., 2020; Geow et al., 2021).

Natural colourants derived from medicinal fruits have a wide range of possible uses. They are used as natural additions in the food industry to improve the nutritional content and aesthetic appeal of products (Henry, 1996). These colourants provide a natural substitute for synthetic colours in lipstick, eyeshadow, and skincare product compositions used in the cosmetics sector (Mahesh, Fathima, & Veena, 2019). Furthermore, the pharmaceutical sector can utilise these colourants' therapeutic qualities to create dietary supplements and pharmaceutical formulations (Rymbai, Sharma, & Srivastav, 2011).

This chapter will explore the extraction of anthocyanins and carotenoids from these fruits found in Iran using different methods, such as aqueous extraction, solvent extraction, acid and alkali extraction, supercritical fluid extraction, ultrasound-assisted extraction, and enzyme-assisted extraction.

3.2 *Medicinal fruits of Iran: an overview*

3.2.1 *Barberry (*Barberis *spp.)*

Native to Iran, barberry is a valuable plant that is grown in abi and poloei varieties (*Berberis integerrima, B. vulgaris*). The red colour of barberries is due to their abundance of anthocyanins and flavonoids (Ardestani et al., 2013). Renowned flavonoid members, anthocyanins have a variety of effects, such as anti-inflammatory, anti-oedema, antioxidant, and anticancer properties. *Berberis integerrima* is a potential treatment for hepatitis, malaria, blood disorders, fever, and gastrointestinal disorders (Farmani et al., 2018).

3.2.2 *Myrtle berries (*Myrtus communis *L.)*

A common medicinal herb in Iranian traditional medicine is *Myrtus communis* L. Phytochemical substances found in myrtle fruit have potential applications in the food sector as well as in medicine. Myrtle berries are mostly composed of flavonoids, anthocyanins (which give myrtle berries their deep blue colour), essential oil, and phenolic substances, including gallic acid. Myrtle's antioxidant activities are attributed to certain compounds, including anthocyanins and polyphenols. As a result of these characteristics, myrtle may find application in dietary supplements and pharmaceutical formulations. As a result, these berries have a neuroprotective effect and can effectively treat hypermenorrhoea and peptic ulcers (Mobli et al., 2016).

3.2.3 *Figs (*Ficus carica*)*

One of the most significant horticultural crops, figs are grown throughout most of Iran. Tea is known to contain the highest concentration of phenolic compounds; nonetheless, fresh and dry figs are rich in proanthocyanin and other phenolic compounds. The primary source of fibre, potassium, calcium, iron, and free sodium is this fruit. Fig fruit also contains high levels of free fat, amino acids, vitamins, carotene, and anthocyanin. These pigments give the fruit its purple and black hue, which highlights its high concentration of antioxidants, polyphenols, anthocyanins, and flavonoids. They also have anti-inflammatory and anticancer effects (Zare & Sotoudeh, 2018).

3.2.4 *Pistachio (*Pistacia atlantica*)*

Pistacia atlantica's fruit and flower are rich sources of flavonoids, resins, inorganic acids, polysaccharides, and phenolic compounds, which have antibacterial properties against both gram-positive and gram-negative bacteria. The fruit and blossom of *P. atlantica* are known to contain a variety of medicinal chemical compounds with strong antibacterial potential (Mostafavi et al., 2021).

3.2.5 *Hawthorn (*Crataegus elbursensis*)*

In the northern reaches of Iran's forests are hawthorn plants with tiny black fruits. For millennia, people have utilised hawthorn, or *Crataegus elbursensis*, for food and medicine. Proanthocyanidins and flavonoids are the source of many of hawthorn's therapeutic qualities. Additionally, several of these flavonoids have antibacterial, anti-inflammatory, and gastroprotective qualities (Sadeghi-Kiakhani et al., 2021).

3.2.6 *Date palms (*Phoenix dactylifera*)*

Iranian natives use the date palm as a multipurpose tree. These have elevated concentrations of carotenoids, especially beta-carotene, an essential precursor of vitamin A. In addition, dates contain flavonoids and anthocyanins, which give them their light to dark brown hue and are well known for their antioxidant properties (Al-Okbi, 2022). The fruit is used as a medication to treat constipation, anaemia, blood purgation, and cough. Date fruit has been shown to improve spleen plaque-forming cell counts, haemagglutinating antibody titres, and the macrophage movement index, which measures cell-mediated immunity (Sadeghi, Valizadeh, & Shermeh, 2015).

*3.2.7 Silverberry (*Elaeagnus angustifolia *L.)*

The Eurasian tree known as Russian olive, or silverberry, has burnt orange skin that is rich in antioxidants, phenols (benzoic acid, cinnamic acid), and flavonoids (myricetin, epigallocatechin gallate). As a result, the fruit of this tree is regarded as a functional food (Patel, 2014). Dried fruit, the pulp powder, and seed of this fruit have been used as an analgesic in Iranian traditional medicine to relieve pain in rheumatoid arthritis. Furthermore, pharmacological research has demonstrated antibacterial, anti-inflammatory, and antinociceptive properties, as well as muscle relaxant action (Mahmoodi et al., 2018).

*3.2.8 Pomegranates (*Punica granatum*)*

Iran holds a significant position in the global pomegranate production and export market. Pomegranates get their rich red colour from anthocyanins, which are abundant in them. Other phenolic chemicals found in these fruits have anti-inflammatory and antioxidant qualities (Bazargani-Gilani, Tajik, & Aliakbarlu, 2014). Pomegranate consumption has been linked to a number of positive health outcomes, including the reduction of blood pressure, inflammation, atherosclerosis, prostate cancer, heart disease, and HIV-1 (Tehranifar et al., 2010).

*3.2.9 Iranian Qara Qara (*Ribes rubrum*)*

The majority of Iran's herbaceous species, including the delicious and sour redcurrant, known as Qara Qara, are found in the Arasbaran vegetation zone, which is extremely rich. Redcurrants are seasoned in the middle of summer. Anthocyanin manifests as a blue pigment in alkaline media and as a red pigment in acidic ones. One of the flavonoids thought to offer the most nutritional benefits to humans throughout their lives is anthocyanin, which has been shown to have positive effects on a number of illnesses, including diabetes, cancer, obesity, neurological disorders, and inflammation (Mollaamin et al., 2021).

3.2.10 *Mulberry (*Morus *spp.)*

One of the main mulberry growers is Iran. For the purpose of producing fruit, three primary mulberry species are described: *Morus alba* L. (white mulberry), *M. rubra* L. (red mulberry), and *M. nigra* L. (black mulberry). These fruits' high anthocyanin concentration makes them excellent providers of phenolic chemicals. It has been demonstrated that phenolics' anti-ageing qualities are due to their antioxidant capabilities. Moreover, it has been proposed that phenolic chemicals may lower the incidence of certain cancers and coronary heart disorders (Farahani et al., 2019).

3.2.11 *Grapes (*Vitis vinifera *spp.)*

One of the major fruit crops in the world, grapes (*Vitis vinifera* L.) include a significant number of various phenolic compounds in their leaves, skins, pulp, and seeds. These compounds may have biologically significant qualities connected to their antioxidant properties. In West Azerbaijan, Iran, there are five indigenous grape kinds grown: Hosseini, Ghara Shira, Agh Shani, Ghara Shani, and Ghara Ghandome. The grapes have different colours, ranging from white to dark red, green, yellow, and black. By neutralising reactive species, the antioxidant chemicals found in them are the primary agents that might greatly lessen their harmful effects and potentially prevent the onset of degenerative processes such as diabetes, cancer, and cardiovascular disease (Farhadi et al., 2016).

3.2.12 *Cornelian cherry (*Cornus mas *L.)*

The plant known as *Cornus mas* L., or cornelian cherry, is native to western Asia, sections of central and southern Europe, and the northern woodlands of Iran. This plant produces berries that are high in anthocyanins, such as pelargonidin-3-galactoside, cyanidin-3-rhamnoglucoside, cyanidin-3-glucoside, and delphinidin-3-glucoside (Demir & Kalyoncu, 2003; Seeram et al., 2002). These anthocyanins improve insulin resistance by causing pancreatic β-cells to secrete more insulin. Consequently, this herb has long been used as a supplement to treat diabetes (Soltani et al., 2015).

3.2.13 *Cherry (*Prunus avium*)*

Due to its delightful flavour and scent, sweet cherries (*Prunus avium* L.), also referred to as wild cherries, are regarded as one of the most prized fruits in the world. Additionally, they are abundant in a variety of nutritional components (such as sugars and organic acids) and phytochemical compounds (such as carotenoids, phenolics, melatonin, and serotonin) that are crucial for biological processes (Fonseca et al., 2021; Gonçalves et al., 2019). Eating cherries is linked to a number of health advantages, including the potential prevention of diabetes, cancer, heart disease, inflammatory illnesses, Alzheimer's disease, and other conditions. Cherries have low glycaemic response, strong antioxidant potential, anti-carcinogenic properties, and inhibition of COX-1 and COX-2 (McCune et al., 2010).

3.2.14 *Jujube (*Ziziphus jujuba *Mill.)*

A strategically important plant, jujube may be grown in many parts of Iran. Jujube fruits possess antioxidants, anti-inflammatory qualities, and antibacterial potential. The compounds in jujube fruits are neuroprotective, anticancer, anti-HIV, sedative-hypnotic, anxiolytic, anti-spastic, anti-fertility/contraceptive, hypotensive and anti-nephritic, cardiotonic, and immune-stimulant, although they can help regulate the regular functions of the gastrointestinal tract and the cardiovascular system (Ghani et al., 2022).

3.2.15 *Saffron (*Crocus sativus*)*

The world's most costly spice, saffron (*Crocus sativus*), is mostly grown in Iran, Spain, Greece, and India. It is renowned for its delicate flavour, bitter taste, and eye-catching yellow colour. Saffron is valued as a medicinal plant in addition to being used as a spice and colouring agent because it contains crocin, picrocrocin, and safranal, which are being researched as novel, safe sources of natural antioxidants for the food industry (Gazeran, Tajalli, & Mohamadi Sani, 2016) (Table 3.1).

3.3 *Techniques used for extraction of colour pigments*

3.3.1 *Aqueous extraction*

Aqueous extraction is a straightforward and eco-friendly method for extracting water-soluble colourants from fruits. This technique uses water as the solvent to dissolve and extract colourants, making it particularly suitable for pigments like anthocyanins, betalains, and some flavonoids that have good solubility in water (Figure 3.1).

Table 3.1 Colour pigment and health benefits of different medicinal fruits found in Iran

Sr. No.	Fruits found in the region	Scientific name of the fruit	Colour of the fruit	Colour pigment present	Health benefits	References
1.	Barberry	*Barberis* spp.	Red	Anthocyanins	Treats gastrointestinal diseases, bleeding, fever, malaria and hepatitis	Ardestani et al. (2013) and Farmani et al. (2018)
2.	Myrtle berries	*Myrtus communis* L	Deep blue	Anthocyanins	Treats hypermenorrhoea and peptic ulcers	Mobli et al. (2016)
3.	Figs	*Ficus carica*	Purple to black	Anthocyanins, carotene	Anti-inflammatory and anticancer effects	Zare and Sotoudeh (2018)
4.	Pistachio	*Pistacia atlantica*	Green-yellow	Flavonoids	Antibacterial properties	Mostafavi et al. (2021)
5.	Hawthorn	*Crataegus elbursensis*	Black	Proanthocyanidin, flavonoids	Antibacterial, anti-inflammatory, and gastroprotective qualities	Sadeghi-Kiakhani et al. (2021)

(*Continued*)

***Table 3.1* (*Continued*)** Colour pigment and health benefits of different medicinal fruits found in Iran

Sr. No.	Fruits found in the region	Scientific name of the fruit	Colour of the fruit	Colour pigment present	Health benefits	References
6.	Date Palms	*Phoenix dactylifera*	Light to dark brown	Beta-carotene	Treats constipation, anaemia, blood purgation, and cough, improves spleen plaque-forming cell counts, haemagglutinating antibody titters, and the macrophage movement index	Al-Okbi (2022) and Sadeghi, Valizadeh, and Shermeh (2015)
7.	Silverberry	*Elaeagnus angustifolia* L.	Burnt orange	Flavonoids (myricetin, epigallocatechin gallate)	Has antibacterial, anti-inflammatory, and antinociceptive properties as well as muscle relaxant action	Patel (2014) and Mahmoodi et al. (2018)
8.	Pomegranates	*Punica granatum*	Deep red	Anthocyanins	Reduces blood pressure, inflammation, atherosclerosis, prostate cancer, heart disease, and HIV-1	Bazargani-Gilani, Tajik, and Aliakbarlu (2014) and Tehranifar et al. (2010)

(*Continued*)

***Table 3.1* (*Continued*)** Colour pigment and health benefits of different medicinal fruits found in Iran

Sr. No.	Fruits found in the region	Scientific name of the fruit	Colour of the fruit	Colour pigment present	Health benefits	References
9.	Iranian Qara	*Ribes rubrum*	Red, blue	Anthocyanins	Lowers diabetes, cancer, obesity, neurological disorders, and inflammation	Mollaamin et al. (2021)
10.	Mulberry	*Morus* spp.	Red, black, white	Anthocyanins	Anti-ageing properties, lowers cancers and coronary heart disorders	Farahani et al. (2019)
11.	Grapes	*Vitis vinifera* spp.	White, dark red, yellow, green, black	Anthocyanins, flavonoids	Reduces diabetes, cancer, and cardiovascular disease	Farhadi et al. (2016)
12.	Cornelian cherry	*Cornus mas* L.	Pink to red, yellow	Anthocyanins (pelargonidin-3-galactoside, cyanidin-3-rhamnoglucoside)	Improves insulin resistance, treats diabetes	Demir and Kalyoncu (2003), Seeram et al. (2002), and Soltani et al. (2015)
13.	Cherry	*Prunus avium*	Bright red to purple	Carotenoids	Prevents diabetes, cancer, heart disease, inflammatory illnesses, Alzheimer's disease, have low glycaemic response, strong antioxidant potential, anti-carcinogenic properties, and inhibits COX 1 and 2	Fonseca et al. (2021), Gonçalves et al. (2019), and McCune et al. (2010)

(*Continued*)

Table 3.1 (Continued) Colour pigment and health benefits of different medicinal fruits found in Iran

Sr. No.	Fruits found in the region	Scientific name of the fruit	Colour of the fruit	Colour pigment present	Health benefits	References
14.	Jujube	*Ziziphus jujuba* Mill.	Green to dark brown	Flavonoids	Neuroprotective, anticancer, anti-HIV, sedative-hypnotic, anxiolytic, anti-spastic, antifertility/contraception, hypotensive and anti-nephritic, cardiotonic, and immune-stimulant, regulate functions of the gastrointestinal tract and the cardiovascular system	Ghani et al. (2022)
15.	Saffron	*Crocus sativus*	Yellow	Carotenoids (crocin, picrocrocin, and safranal)	Antioxidant properties	Gazeran, Tajalli, and Mohamadi Sani (2016)

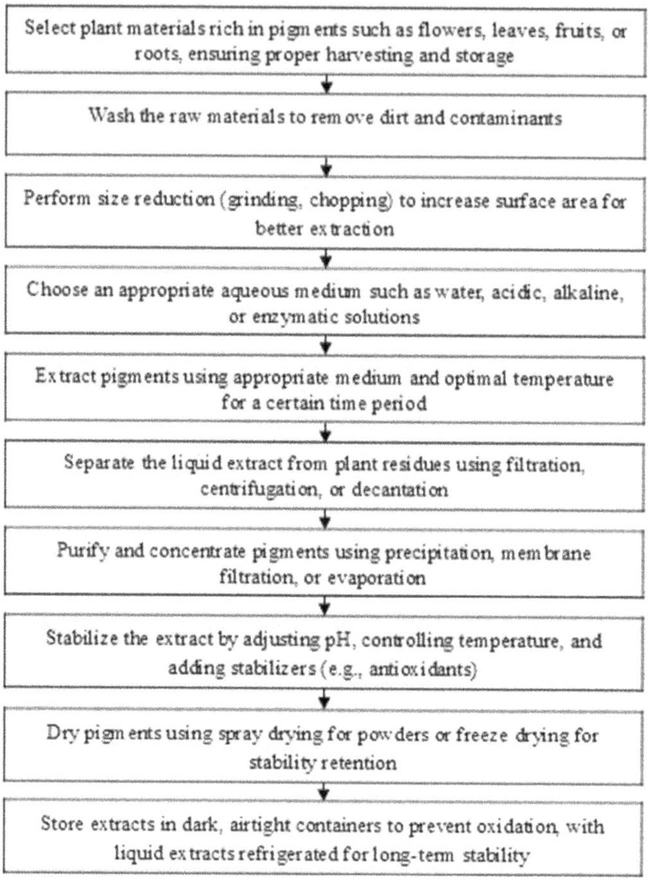

Figure 3.1 Aqueous extraction of colour pigments.

3.3.1.1 Procedure

See Figure 3.1.

3.3.1.2 Key advantages

- **Non-Toxic:** Water is a safe and non-toxic solvent, making the extracts suitable for food, cosmetics, and pharmaceutical applications.
- **Cost-Effective:** Water is inexpensive and readily available.
- **Eco-Friendly:** The process is environmentally friendly, producing minimal waste and avoiding the use of harmful organic solvents.

3.3.1.3 Application

Aqueous extraction is suitable for extracting water-soluble colourants like anthocyanins from figs (Dueñas et al., 2008).

3.3.2 Solvent extraction

Solvent extraction is a widely used method for isolating colourants from fruits, especially when the target compounds are not easily extracted using water. This method uses organic solvents to dissolve the colourants, which are then separated from the fruit material. The choice of solvent depends on the solubility of the specific colourants you are targeting, such as anthocyanins, carotenoids, or flavonoids (Figure 3.2).

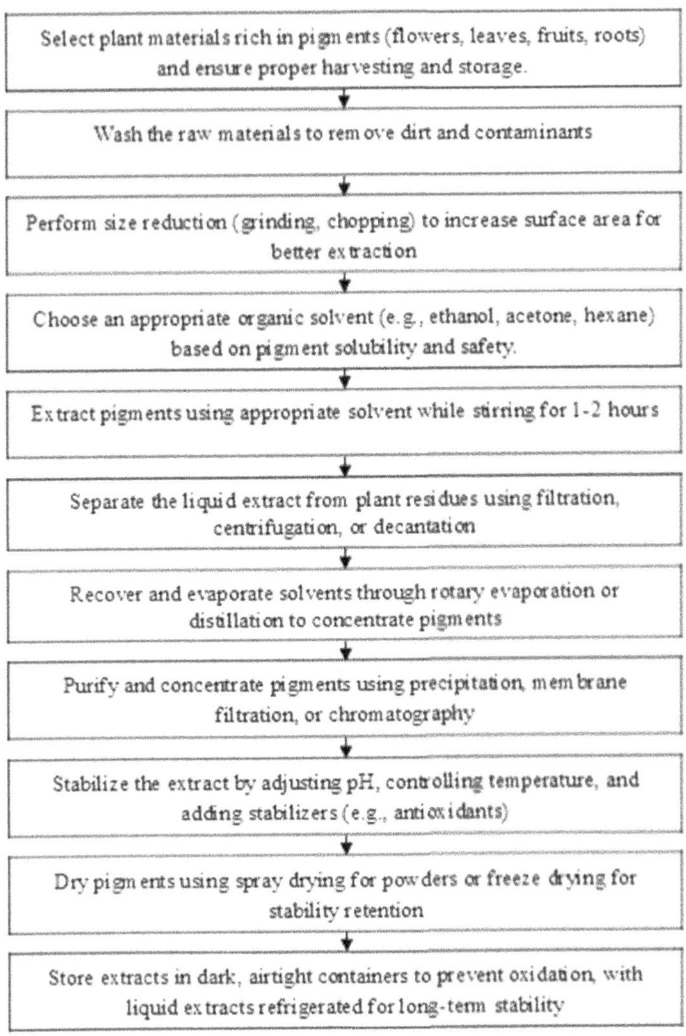

Figure 3.2 Solvent extraction of colour pigments.

3.3.2.1 *Procedure*

See Figure 3.2.

3.3.2.2 *Key advantages*

- **High Efficiency:** Effective at extracting both polar and non-polar colourants, depending on the solvent used.
- **Flexibility:** Can be tailored to target specific colourants by selecting the appropriate solvent.
- **Scalability:** Suitable for both small-scale laboratory experiments and large-scale industrial processes.

3.3.2.3 *Application*

Solvent extraction is effective for both anthocyanins and carotenoids. For example, ethanol can extract carotenoids from date palms (El-Rayes, 2009) and anthocyanins from berries and pomegranate (Zam, Ali, & Ibrahim, 2017; Kulkarni et al., 2011).

3.3.3 *Reflux extraction*

Reflux extraction is a technique used to extract colourants from fruits and other natural materials by continuously boiling the solvent and condensing the vapour back into the mixture. This method allows for efficient extraction by maintaining a constant temperature and solvent volume, which ensures maximum contact between the solvent and the fruit material (Figure 3.3).

3.3.3.1 *Procedure*

See Figure 3.3.

3.3.3.2 *Key advantages*

- **Efficient Extraction:** Maintains a consistent temperature and solvent volume, enhancing extraction efficiency.
- **Continuous Process:** The solvent is continuously recycled, reducing the need for additional solvent.
- **Suitable for Heat-Stable Compounds:** Ideal for extracting colourants that are stable at elevated temperatures.

3.3.3.3 *Application*

Reflux extraction is suitable for extracting a wide range of colourants from fruits such as barberries, including anthocyanins, carotenoids, and flavonoids. The method is particularly effective for compounds that are heat-stable and can withstand prolonged exposure to the extraction solvent (Sharifi & Hassani, 2012).

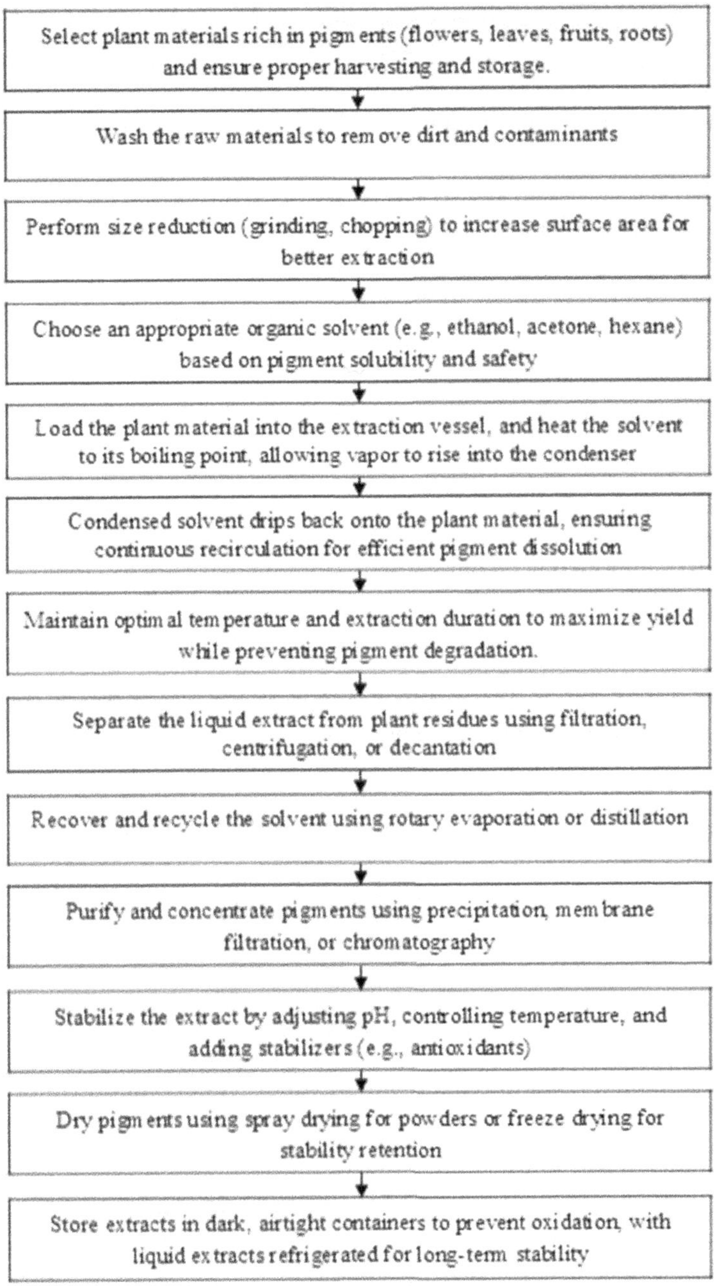

Figure 3.3 Reflux extraction of colour pigments.

3.3.4 *Supercritical fluid extraction (SFE)*

Supercritical fluid extraction (SFE) is a sophisticated method used to extract colourants and other bioactive compounds from fruits and other plant materials. The process uses a supercritical fluid, typically carbon dioxide (CO_2), which has properties of both a gas and a liquid, allowing it to penetrate materials like a gas but dissolve substances like a liquid. This method is particularly effective for extracting non-polar compounds, such as carotenoids and certain flavonoids, but it can be modified to extract a wide range of colourants by using co-solvents (Figure 3.4).

3.3.4.1 *Procedure*

See Figure 3.4.

3.3.4.2 *Key advantages*

- **Selective Extraction:** SFE allows for the selective extraction of compounds by adjusting pressure and temperature.
- **Mild Conditions:** The process operates at relatively low temperatures, preserving heat-sensitive compounds.
- **No Residual Solvents:** CO_2 evaporates completely, leaving no solvent residues in the extract.
- **Environmentally Friendly:** CO_2 is non-toxic, non-flammable, and can be recycled within the process.

3.3.4.3 *Application*

This method is particularly suitable for extracting high-value non-polar colourants like carotenoids (e.g., β-carotene, lycopene), anthocyanins, and other lipophilic pigments from saffron and date palms (Heydari & Haghayegh, 2014; Ghafoor et al., 2022).

3.3.5 *Ultrasound-assisted extraction (UAE)*

Ultrasound-assisted extraction (UAE) is an innovative technique that utilises ultrasonic waves to enhance the extraction of colourants and other bioactive compounds from fruits and other plant materials. The method involves the use of high-frequency sound waves to create cavitation bubbles in the solvent. When these bubbles collapse, they generate intense localised pressure and temperature, disrupting the cell walls of the fruit material and facilitating the release of the target compounds into the solvent (Figure 3.5).

3.3.5.1 *Procedure*

See Figure 3.5.

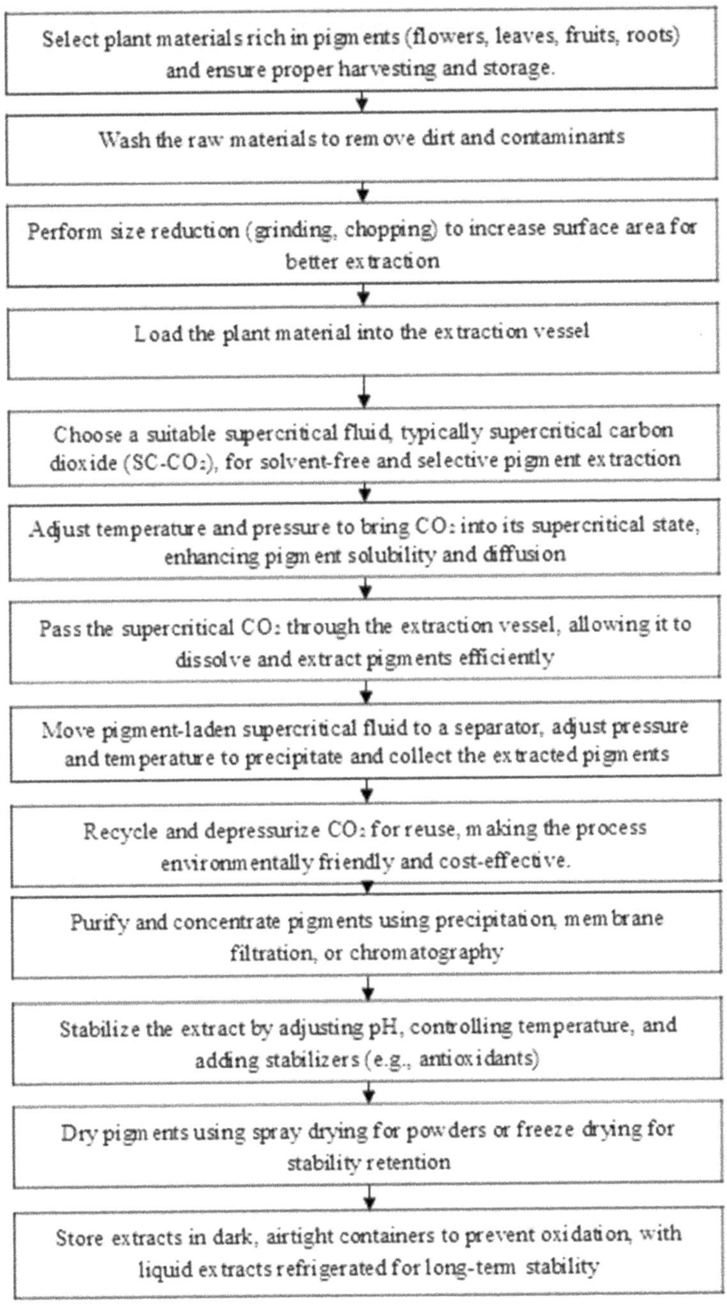

Figure 3.4 Supercritical fluid extraction of colour pigments.

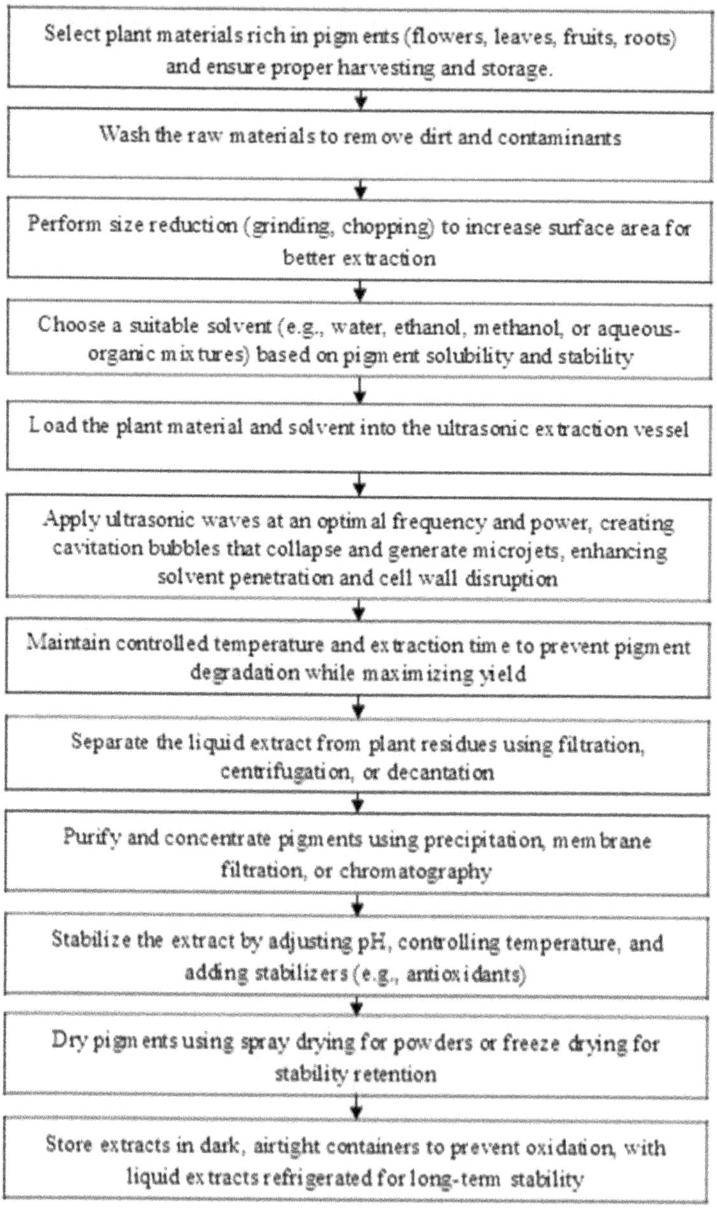

Figure 3.5 Ultrasound-assisted extraction of colour pigments.

3.3.5.2 Key advantages of UAE

- **Increased Efficiency:** Enhanced extraction yields with reduced solvent usage and shorter extraction times.
- **Mild Conditions:** Lower temperatures can be used, making it suitable for heat-sensitive compounds.
- **Versatility:** Effective for a wide range of solvents and plant materials.
- **Scalability:** Can be adapted for both small-scale laboratory use and large-scale industrial applications.

3.3.5.3 Application

Ultrasound-assisted extraction is widely used for the extraction of natural colourants, antioxidants, and other bioactive compounds from fruits, vegetables, and herbs. It is particularly effective for extracting compounds like anthocyanins, carotenoids, and polyphenols from fruit materials like pomegranate, mulberries, and hawthorn (Sharayei et al., 2019; Espada-Bellido et al., 2017; Sadeghi-Kiakhani et al., 2021).

3.3.6 Enzyme-assisted extraction (EAE)

Enzyme-assisted extraction (EAE) is a method that utilises specific enzymes to facilitate the extraction of colourants and other bioactive compounds from fruits. Enzymes break down the cell walls and cellular structures, making it easier to release and extract target compounds. This technique can be highly effective in obtaining high-quality extracts with minimal degradation of sensitive components (Figure 3.6).

3.3.6.1 Procedure

See Figure 3.6.

3.3.6.2 Key advantages of EAE

- **Increased Efficiency:** Enzymes enhance the release of colourants by breaking down cell walls and structures, leading to higher extraction yields.
- **Reduced Processing Time:** EAE can shorten the extraction time compared to traditional methods.
- **Improved Quality:** Enzyme-assisted extraction can reduce the degradation of sensitive compounds, resulting in higher-quality extracts.
- **Lower Solvent Usage:** By improving extraction efficiency, EAE may reduce the amount of solvent needed.

3.3.6.3 Application

EAE is particularly effective for extracting natural colourants such as anthocyanins, carotenoids, and flavonoids from fruits with high pectin

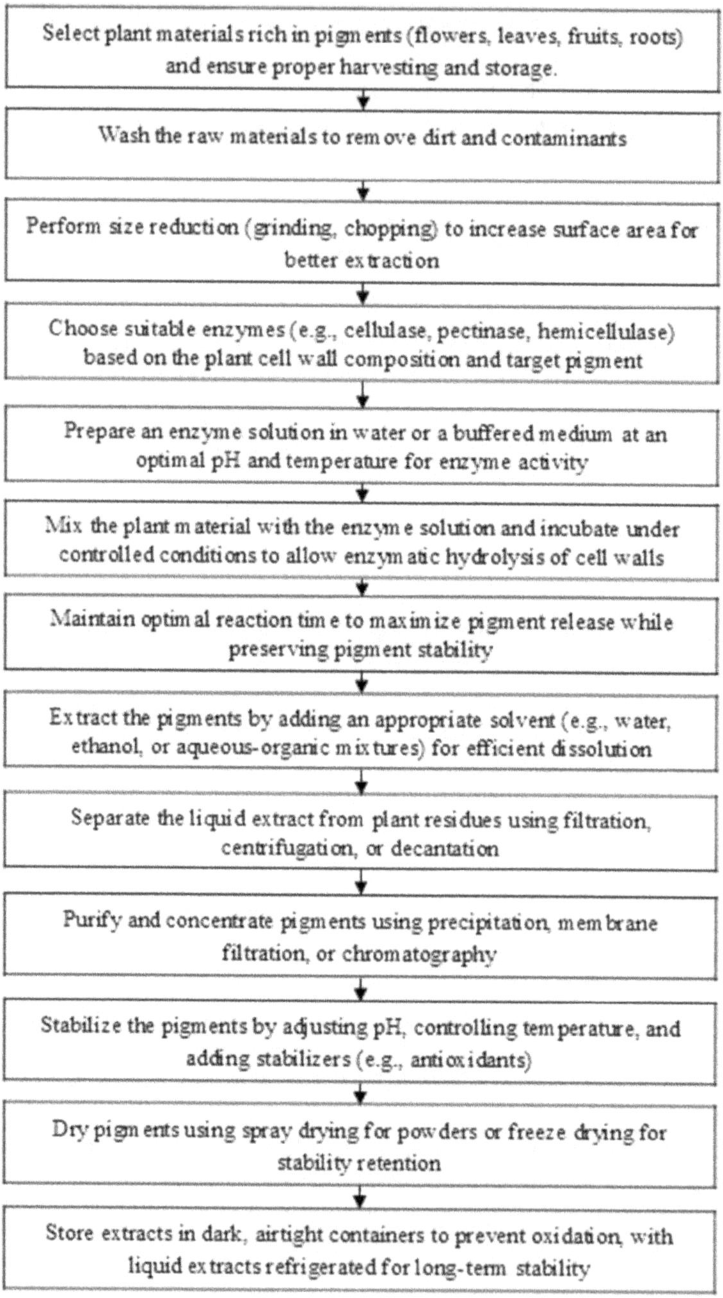

Figure 3.6 Enzyme-assisted extraction of colour pigments.

content, such as saffron, grapes, and mulberries (Lotfi et al., 2015; Maier et al., 2008; Zhou et al., 2021).

3.3.7 Microwave-assisted extraction (MAE)

Microwave-assisted extraction (MAE) is an efficient technique for extracting colourants and other bioactive compounds from fruits and plants. This method utilises microwave energy to heat the solvent and plant material, accelerating the extraction process. MAE is particularly advantageous because it reduces extraction time, improves yield, and often requires less solvent compared to traditional methods (Figure 3.7).

3.3.7.1 Procedure

See Figure 3.7.

3.3.7.2 Key advantages of MAE

- **Rapid Extraction:** Significantly reduces extraction time compared to traditional methods.
- **Higher Yield:** Often results in higher extraction yields due to efficient cell wall disruption.
- **Reduced Solvent Use:** Requires less solvent, making the process more environmentally friendly.
- **Energy Efficiency:** Microwaves selectively heat the solvent and plant material, reducing overall energy consumption.

3.3.7.3 Application

Microwave-assisted extraction is used to extract natural colourants like anthocyanins, flavonoids, and carotenoids from fruits. MAE is particularly effective for extracting colourants from fruits like pomegranate, which contain water-soluble pigments (Zahed, Esmaeilzadeh Kenari, & Farahmandfar, 2023) (Table 3.2).

3.4 Applications of colourants extracted

Natural colourants extracted from these medicinal fruits have a wide range of applications across various fields, thanks to their bioactive properties and aesthetic value. Some notable applications include:

1. **Pharmaceutical Industry:**
    - **Natural Colourants in Tablets and Capsules:** Colourants are used to enhance the visual appeal of pharmaceuticals and can be derived from medicinal fruits (Naji-Tabasi et al., 2021).
    - **Active Ingredients:** Some colourants have therapeutic properties and are used as active ingredients in medicinal formulations. For

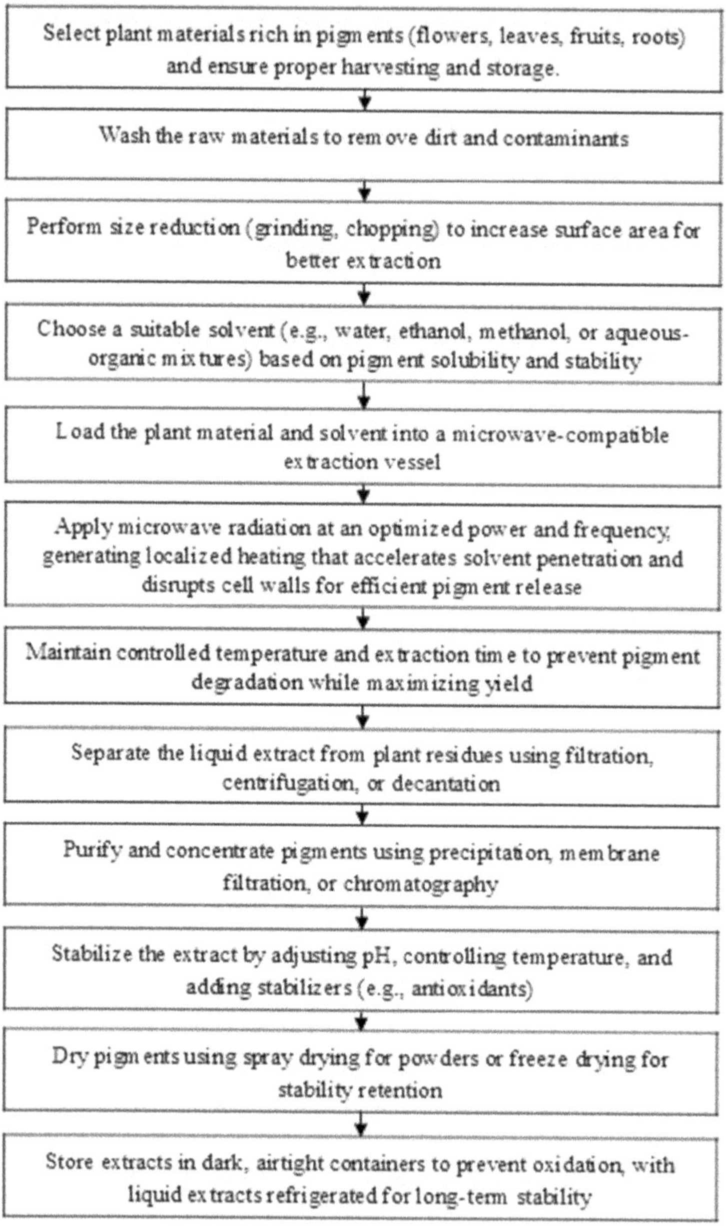

Figure 3.7 Microwave-assisted extraction of colour pigments.

Table 3.2 Different extraction methods for colour extraction from various medicinal fruits found in Iran

Sr. No.	Extraction technique	Solvents used	Fruits used	Advantages of technique used	References
1.	Aqueous extraction	Water	Figs	Non-toxic, cost-effective	Dueñas et al. (2008)
2.	Solvent extraction	Ethanol, water, methanol, acetone, ethyl acetate, hexane	Date palm, berries, pomegranate	Flexible as per colourant, more efficient	El-Rayes (2009), Zam, Ali, and Ibrahim (2017), and Kulkarni et al. (2011)
3.	Reflux extraction	Ethanol, methanol, water, ethyl acetate	Barberries	Requires less solvent, can extract elevated temperature compounds	Sharifi and Hassani (2012)
4.	Supercritical fluid extraction (SFE)	Supercritical CO_2	Saffron, date palm	Environment friendly, no residual solvent	Heydari and Haghayegh (2014) and Ghafoor et al. (2022)
5.	Ultrasound-assisted extraction (UAE)	Ethanol, methanol, water, ethyl acetate	Pomegranate, mulberries, hawthorn	Can be used for heat-sensitive compounds, versatile	Sharayei et al. (2019), Espada-Bellido et al. (2017), and Sadeghi-Kiakhani et al. (2021)
6.	Enzyme-assisted extraction (EAE)	Water, ethanol, methanol, acetone, ethyl acetate	Saffron, grapes, mulberries	Improved quality, lower solvent requirement, more efficient	Lotfi et al. (2015), Maier et al. (2008), and Zhou et al. (2021)
7.	Microwave-assisted extraction (MAE)	Water, ethanol, methanol, acetone, ethyl acetate	Pomegranate	Lower solvent requirement, rapid extraction, higher yield	Zahed, Esmaeilzadeh Kenari, and Farahmandfar (2023)

example, anthocyanins from fruits like elderberries may have antioxidant and anti-inflammatory effects (Jafarpour et al., 2016).

2. **Cosmetics and Personal Care Products:**
    - **Natural Dyes in Cosmetics:** Colourants from medicinal fruits are used in products like lipsticks, shampoos, and lotions for their natural colours and potential skin benefits (Pasdaran et al., 2023).
    - **Skin Care:** Some colourants, such as those from pomegranate or acai berries, are used in skincare products for their antioxidant properties (Aafi et al., 2022).
3. **Food and Beverage Industry:**
    - **Natural Food Colouring:** Extracted colourants are used to provide natural colouring to foods and beverages, such as juices, jams, and confectioneries (Asadnejad et al., 2018).
    - **Health Supplements:** Some fruit colourants are included in health supplements due to their high antioxidant content and potential health benefits (Shakeri et al., 2018).
    - **Active Packaging:** Some natural colourants from fruits are used in active or smart packaging for food materials that enhances the quality and safety of food product (Abedi-Firoozjah et al., 2023).
4. **Textile Industry:**
    - **Natural Dyes for Fabrics:** Colourants from medicinal fruits are used in the textile industry to dye fabrics and fibres. These natural dyes are often preferred for their eco-friendliness, sustainability, and the unique, vibrant colours they provide (Kiumarsi & Gashti, 2015; Barani, Rezaee, & Maleki, 2019; Sadeghi-Kiakhani et al., 2021).
5. **Research and Development:**
    - **Bioactive Compound Studies:** Extracted colourants are studied for their potential health benefits, including their antioxidant, anti-inflammatory, and antimicrobial properties (Asadi-Gharneh, Mohammadzamani, & Karimi, 2017; Nobari et al., 2022).
    - **Formulation Development:** Researchers use fruit colourants to develop new formulations for various applications, including cosmetics, pharmaceuticals, and food products (Mahdavi et al., 2016).
    - **Biological Stains:** Natural colourants extracted are used for biological staining in various experiments (Bathaie, Farajzade, & Hoshyar, 2014).
6. **Traditional Medicine and Herbal Products:**
    - **Herbal Remedies:** In traditional medicine, colourants from medicinal fruits may be used to enhance the visual appeal and perceived efficacy of herbal products and remedies (Ghafari, Fahimi, & Sahranavard, 2017).

These applications leverage both the aesthetic and potential health benefits of colourants extracted from medicinal fruits, making them valuable in a range of industries (Table 3.3).

Table 3.3 Applications of different colours extracted from various medicinal fruits of Iran

Sr. No.	Fruit found in the country	Pigment extracted	Colour imparted	Applications	References
1.	Pomegranate (*Punica granatum*)	Anthocyanins, ellagitannins, and punicalagins	Deep red to purple	**In Pharmaceutical Industry:** reduce oxidative stress and inflammation, support cardiovascular health, improve blood flow, reduce the risk of certain cancers	Fateh et al. (2013)
				In Cosmetic and Personal Care: has skin-firming, anti-wrinkle, brightening properties and antioxidant effects, gives natural hue in lip balms	Dhineshkumar and Ramasamy (2016)
				In Food and Beverage Industry: as natural colourant in fruit drinks, health juices, sorbets, confectioneries, ice creams	Kandylis and Kokkinomagoulos (2020)
				In Textile Industry: used to dye cotton, wool, silk fabrics in soft yellow to brown hues, as a mordant in combination with other natural dyes to improve colour fastness and create a range of shades	Adeel et al. (2009)
				In Research and Development: in developing health supplements and functional foods, as natural preservatives to extend shelf life in food products	Das et al. (2021)

(*Continued*)

Table 3.3 (Continued) Applications of different colours extracted from various medicinal fruits of Iran

Sr. No.	Fruit found in the country	Pigment extracted	Colour imparted	Applications	References
2.	Barberry (*Berberis vulgaris*)	Berberine and anthocyanins	Yellow	**In Pharmaceutical Industry:** has antimicrobial, anti-inflammatory, and antidiabetic properties, supports heart health and weight management	Barani, Rezaee, and Maleki (2019)
				In Cosmetic and Personal Care: used in skin-brightening by inhibiting melanin production and reduce hyperpigmentation, soothe the scalp by using in shampoo and conditioners	Hassan et al. (2023)
				In Food and Beverage Industry: as natural colourant in rice dishes, beer, cakes, cookies, sauces and dressings, syrups	Lemoine et al. (2021)
				In Textile Industry: used to dye silk and wool in bright yellow hues, in printing on fabrics, particularly for creating intricate designs	Motaghi (2018)
				In Research and Development: for developing new drugs for its antimicrobial, anti-inflammatory, and potential anticancer properties, as a nutraceutical	Stabnikova, Stabnikov, and Paredes-López (2024)

(Continued)

Table 3.3 **(*Continued*)** Applications of different colours extracted from various medicinal fruits of Iran

Sr. No.	Fruit found in the country	Pigment extracted	Colour imparted	Applications	References
3.	Saffron (*Crocus sativus*)	Crocin and crocetin	Deep orange-red	**In Pharmaceutical Industry:** has antioxidant effects, mood-enhancing properties, helps in treating depression, Alzheimer's disease, and eye health	Moshiri, Vahabzadeh, and Hosseinzadeh (2015)
				In Cosmetic and Personal Care: has brightening effects, helps to reduce dark spots and even out skin tone, has anti-ageing effects	Fekrat (2003)
				In Food and Beverage Industry: as natural colourant in savoury dishes, teas, baked goods, soups, sauces	Fujii et al. (2022)
				In Textile Industry: used to dye silk fabrics in luxurious golden hues, or embellishments or accent colours in textile art	Ali et al. (2014)
				In Research and Development: in treating neurodegenerative diseases like Alzheimer's, in inhibiting cancer cell growth	Ezati et al. (2023)

(*Continued*)

Table 3.3 (Continued) Applications of different colours extracted from various medicinal fruits of Iran

Sr. No.	Fruit found in the country	Pigment extracted	Colour imparted	Applications	References
4.	Mulberry (*Morus* spp.)	Anthocyanins and resveratrol	Dark purple to blue	**In Pharmaceutical Industry:** has antioxidant and anti-inflammatory effects. Supports longevity and cardiovascular health	Acharya, Bagchi, and Gangopadhyay (2022)
				In Cosmetic and Personal Care: used in natural hair dyes, lip and cheek tints, improve skin elasticity and reduce the signs of ageing	Thamrongwatwongsa et al. (2022)
				In Food and Beverage Industry: as natural colourant in alcoholic beverages, smoothies, tarts, cakes, jams, syrups, preserves	Kucelova et al. (2016)
				In Textile Industry: used to dye cotton and silk fabrics in shades of purple and deep red, in traditional tie-dyeing techniques	Wang, Hussain, and Zhou (2017)
				In Research and Development: to regulate blood glucose levels in diabetes, for sustainable textile dyeing	Sheng et al. (2014)

(Continued)

***Table 3.3* (*Continued*)** Applications of different colours extracted from various medicinal fruits of Iran

Sr. No.	Fruit found in the country	Pigment extracted	Colour imparted	Applications	References
5.	Black currant (*Ribes nigrum*)	Anthocyanins, particularly delphinidin and cyanidin	Dark purple to black	**In Pharmaceutical Industry:** strong antioxidants, support eye health, improve immune function, and protect against cardiovascular diseases	Karjalainen et al. (2008)
				In Cosmetic and Personal Care: in creams and lotions for its anti-inflammatory properties, in eyeliners, lipsticks for natural colour	Petrov Ivanković et al. (2023)
				In Food and Beverage Industry: as natural colourant in wines, cordial, muffins, pies, ice creams, smoothies	Xue et al. (2022)
				In Textile Industry: used to dye wool and linen fabrics in deep purple and black hues, in textile art and custom fabric creations	Yasukawa et al. (2017)
				In Research and Development: to improve cardiovascular health, enhance immune function	Ambrożewicz et al. (2013)

(*Continued*)

Table 3.3 **(*Continued*)** Applications of different colours extracted from various medicinal fruits of Iran

Sr. No.	Fruit found in the country	Pigment extracted	Colour imparted	Applications	References
6.	Cherries (*Prunus avium*, *Prunus cerasus*)	Anthocyanins	Bright red, purple	**In Pharmaceutical Industry:** reduces inflammation, pain relief (especially in conditions like arthritis), and improves sleep quality due to their role in melatonin production	Dinda et al. (2016)
				In Cosmetic and Personal Care: hydrating and antioxidant effects, helping to refresh and protect the skin	Brudzyńska, Sionkowska, and Grisel (2021)
				In Food and Beverage Industry: as natural colourant in food drinks, cocktails, flavoured water, gummies, candies, sauces and condiments	Ghosh et al. (2022)
				In Textile Industry: used to dye silk and cotton in bright red and burgundy shades, in fabric printing to create colourful patterns	Brudzyńska, Sionkowska, and Grisel (2021)
				In Research and Development: in developing supplements for arthritis and other inflammatory conditions, sleep-promoting products	Milea et al. (2019)

(*Continued*)

Table 3.3 **(*Continued*)** Applications of different colours extracted from various medicinal fruits of Iran

Sr. No.	Fruit found in the country	Pigment extracted	Colour imparted	Applications	References
7.	Grapes (*Vitis vinifera*)	Anthocyanins and resveratrol	Black, red	**In Pharmaceutical Industry:** are antioxidants that support heart health, improve blood circulation, may have anti-ageing effects	Khoo et al. (2017)
				In Cosmetic and Personal Care: used in anti-ageing creams, shampoos and conditioners to strengthen hair and enhance its natural colour	Leal et al. (2020)
				In Food and Beverage Industry: as natural colourant in grape juice, wine, breads, pastries, confectioneries	Aberoumand (2011)
				In Textile Industry: used to dye cotton and wool in rich red and purple hues	Iqbal et al. (2022)
				In Research and Development: in cardioprotective effects, enhancing the colour stability and antioxidant properties of grape-derived products in wine and beverages	Ananga et al. (2013)

(*Continued*)

Table 3.3 **(*Continued*)** Applications of different colours extracted from various medicinal fruits of Iran

Sr. No.	Fruit found in the country	Pigment extracted	Colour imparted	Applications	References
8.	Quince (*Cydonia oblonga*)	Carotenoids	Yellow to orange	**In Pharmaceutical Industry:** are antioxidants, maintains vision and may help prevent age-related macular degeneration, support digestive health and help regulate bowel movements	Najman et al. (2023)
				In Cosmetic and Personal Care: in soothing creams for sensitive or irritated skin, in hair masks to nourish the scalp and improve hair texture	Lorenz et al. (2008)
				In Food and Beverage Industry: as natural colourant in preserves, sauces, pies, tarts	Wojdyło et al. (2013)
				In Textile Industry: used to dye linen and silk fabrics in soft yellow to golden hues, in pastel dyeing processes	Danila et al. (2021)
				In Research and Development: antioxidant capacity in development of health supplements	Nawirska-Olszańska et al. (2011)

(*Continued*)

***Table* 3.3 (*Continued*)** Applications of different colours extracted from various medicinal fruits of Iran

Sr. No.	Fruit found in the country	Pigment extracted	Colour imparted	Applications	References
9.	Apricot (*Prunus armeniaca*)	Carotenoids (beta-carotene)	Orange	**In Pharmaceutical Industry:** have strong antioxidant properties, promotes healthy skin, supports eye health by filtering harmful blue light and reducing the risk of macular degeneration	Al-Soufi et al. (2022)
				In Cosmetic and Personal Care: in tinted moisturisers for natural warm tint, in sunscreen for protection against UV radiations	Arora et al. (2019)
				In Food and Beverage Industry: as natural colourant in nectars, smoothies, cocktails, spreads, dried fruit mixes, tarts	Kasapoğlu, Kahraman, and Törnük (2020)
				In Textile Industry: used to dye cotton and silk fabrics in warm orange and yellow shades, in fabric painting	Popescu (2024)
				In Research and Development: in promoting eye health, food fortification	Aziz et al. (2020)

(*Continued*)

Table 3.3 (Continued) Applications of different colours extracted from various medicinal fruits of Iran

Sr. No.	Fruit found in the country	Pigment extracted	Colour imparted	Applications	References
10.	Persian Walnut (*Juglans regia*)	Juglone and polyphenols	Brown	**In Pharmaceutical Industry:** have antioxidant and anti-inflammatory properties, supporting brain and heart health	Sadighara et al. (2016)
				In Cosmetic and Personal Care: in hair dyes for a natural brown colour, in natural tanning products	Beiki, Najafpour, and Hosseini (2018)
				In Food and Beverage Industry: as natural colourant in cakes, cookies, caramel, stuffings and sauces	Habibi et al. (2023)
				In Textile Industry: used to dye wool and cotton fabrics in rich brown tones, give fabrics an antique or vintage look	Mirshahi et al. (2013)
				In Research and Development: in development of dietary supplements, a sustainable and eco-friendly dye for textiles, focusing on improving dye uptake and fastness	Sun et al. (2019)

3.5 Conclusion

Investigating natural colourants made from Iranian medicinal fruits offers an appealing option towards the creation of long-lasting, health-promoting substitutes for artificial dyes. The rich history of fruits like *barberry, pomegranate,* and *jujube*—which have been prized for their colourful pigments as well as their medical qualities—has been brought to light in this chapter. It has been established by a thorough analysis of numerous extraction methods that these fruits are reliable suppliers of natural colourants with enormous industrial application potential.

Anthocyanins and flavonoids, which contribute to the colours but also provide antioxidant effects, are among the bioactive components found in the extracted colourants, according to a chemical study. These colourants' capacity to remain stable in a variety of environments highlights how useful they are for applications in the food, cosmetics, and pharmaceutical industries.

Despite the promising potential, challenges remain, particularly in scaling up production, reducing costs, and navigating regulatory hurdles. However, with increasing consumer demand for natural and eco-friendly products, the prospects for the commercialisation of these natural colourants are strong.

In summary, the process of extracting natural colourants from Iranian medicinal fruits is an example of how modern scientific innovation and traditional knowledge have come together. By utilising these natural resources, it will be possible to produce goods that are both culturally and environmentally significant, helping to preserve Iran's rich natural history. Prolonged investigation and growth in this domain may result in noteworthy progress in the natural colourant sector as well as the wider framework of sustainable development.

References

Aafi, E., Shams Ardakani, M. R., Ahmad Nasrollahi, S., Mirabzadeh Ardakani, M., Samadi, A., Hajimahmoodi, M., ... & Firooz, A. (2022). Brightening effect of Ziziphus jujuba (jujube) fruit extract on facial skin: A randomized, double-blind, clinical study. *Dermatologic Therapy, 35*(7), e15535.

Abedi-Firoozjah, R., Parandi, E., Heydari, M., Kolahdouz-Nasiri, A., Bahraminejad, M., Mohammadi, R., ... & Garavand, F. (2023). Betalains as promising natural colorants in smart/active food packaging. *Food Chemistry, 424*, 136408.

Aberoumand, A. (2011). A review article on edible pigments properties and sources as natural biocolorants in foodstuff and food industry. *World Journal of Dairy & Food Sciences, 6*(1), 71–78.

Acharya, R., Bagchi, T., & Gangopadhyay, D. (2022). Mulberry as a valuable resource for food and pharmaceutical industries: A review. In Kumar, S. (Ed.), *Medicinal Plants*. London: IntechOpen. http://dx.doi.org/10.5772/intechopen.104631

Adeel, S., Ali, S., Bhatti, I. A., & Zsila, F. (2009). Dyeing of cotton fabric using pomegranate (Punica granatum) aqueous extract. *Asian Journal of Chemistry, 21*(5), 3493.

Ahlawat, T., Patel, N. L., Agnihotri, R., Patel, C. R., & Tandel, Y. (2016). Black mulberry (Morus nigra). *Underutilized Fruit Crops: Importance and Cultivation, 195*, 212.

Ali, N. F., El-Khatib, E. M., El-Mohamedy, R. S. S., & Ramadan, M. A. (2014). Antimicrobial activity of silk fabrics dyed with saffron dye using microwave heating. *International Journal of Current Microbiology and Applied Sciences, 3*(10), 140–146.

Al-Okbi, S. Y. (2022). Date palm as source of nutraceuticals for health promotion: A review. *Current Nutrition Reports, 11*(4), 574–591.

Al-Soufi, M. H., Alshwyeh, H. A., Alqahtani, H., Al-Zuwaid, S. K., Al-Ahmed, F. O., Al-Abdulaziz, F. T., ... & Mediani, A. (2022). A review with updated perspectives on nutritional and therapeutic benefits of apricot and the industrial application of its underutilized parts. *Molecules, 27*(15), 5016.

Ambrożewicz, E., Augustyniak, A., Gęgotek, A., Bielawska, K., & Skrzydlewska, E. (2013). Black-currant protection against oxidative stress formation. *Journal of Toxicology and Environmental Health, Part A, 76*(23), 1293–1306.

Ananga, A., Georgiev, V., Ochieng, J., Phills, B., & Tsolova, V. (2013). Production of anthocyanins in grape cell cultures: A potential source of raw material for pharmaceutical, food, and cosmetic industries. In Poljuha, D., & Sladonja, B. (Eds.), *The Mediterranean Genetic Code- Grapevine and Olive, 1* (pp. 247–287). London, UK: InTech.

Ardestani, S. B., Sahari, M. A., Barzegar, M., & Abbasi, S. (2013). Some physicochemical properties of Iranian native barberry fruits (abi and poloei): Berberis integerrima and Berberis vulgaris. *Journal of Food and Pharmaceutical Sciences, 1*(3).

Arora, R., Aggarwal, G., Dhingra, G. A., & Nagpal, M. (2019). Herbal active ingredients used in skin cosmetics. *Asian Journal of Pharmaceutical and Clinical Research, 12*(9), 7–15.

Asadi-Gharneh, H. A., Mohammadzamani, M., & Karimi, S. (2017). Evaluation of physico-chemical properties and bioactive compounds of some Iranian pomegranate cultivars. *International Journal of Fruit Science, 17*(2), 175–187.

Asadnejad, S., Nabizadeh, R., Nazarinia, A., Jahed, G. R., & Alimohammadi, M. (2018). Data on prevalence of additive colors in local food and beverage products, Tehran, Iran. *Data in Brief, 19*, 2104–2108.

Aseel, D. G., Behiry, S. I., & Abdelkhalek, A. (2023). Green and cost-effective nanomaterials synthesis from desert plants and their applications. In Husen, A. (Ed.), *Secondary Metabolites Based Green Synthesis of Nanomaterials and Their Applications* (pp. 327–357). Singapore: Springer Nature Singapore.

Aziz, M., Yasmin, I., Batool, R., Khan, W. A., Naz, S., Ashraf, F., ... & Iqbal, R. (2020). Exploring the effect of apricot addition on nutritional, antioxidant, textural and sensory characteristics of cookies apricot supplemented functional cookies. *Italian Journal of Food Science, 32*(4). https://doi.org/10.14674/IJFS.1832

Barani, H., Rezaee, K., & Maleki, H. (2019). Influence of dyeing conditions of natural dye extracted from Berberis integerrima fruit on color shade of woolen yarn. *Journal of Natural Fibers, 16*(4), 524–535.

Bathaie, S. Z., Farajzade, A., & Hoshyar, R. (2014). A review of the chemistry and uses of crocins and crocetin, the carotenoid natural dyes in saffron, with particular emphasis on applications as colorants including their use as biological stains. *Biotechnic & Histochemistry*, *89*(6), 401–411.

Bazargani-Gilani, B., Tajik, H., & Aliakbarlu, J. (2014). Physicochemical and antioxidative characteristics of Iranian pomegranate (Punica granatum L. cv. Rabbab-e-Neyriz) juice and comparison of its antioxidative activity with Zataria multiflora Boiss essential oil. *Veterinary Research Forum*, *5*(4), p. 313. Urmia, Iran: Faculty of Veterinary Medicine, Urmia University.

Beiki, T., Najafpour, G. D., & Hosseini, M. (2018). Evaluation of antimicrobial and dyeing properties of walnut (Juglans regia L.) green husk extract for cosmetics. *Coloration Technology*, *134*(1), 71–81.

Brudzyńska, P., Sionkowska, A., & Grisel, M. (2021). Plant-derived colorants for food, cosmetic and textile industries: A review. *Materials*, *14*(13), 3484.

Chemat, F., Vian, M. A., Fabiano-Tixier, A. S., Nutrizio, M., Jambrak, A. R., Munekata, P. E., … & Cravotto, G. (2020). A review of sustainable and intensified techniques for extraction of food and natural products. *Green Chemistry*, *22*(8), 2325–2353.

Danila, A., Muresan, E. I., Chirila, L., & Coroblea, M. (2021). Natural dyes used in textiles: A review. *The 7th International Symposium TTPF*, Iasi, Romania (pp. 52–59).

Das, A. K., Nanda, P. K., Chowdhury, N. R., Dandapat, P., Gagaoua, M., Chauhan, P., … & Lorenzo, J. M. (2021). Application of pomegranate by-products in muscle foods: Oxidative indices, colour stability, shelf life and health benefits. *Molecules*, *26*(2), 467.

Demir, F., & Kalyoncu, I. H. (2003). Some nutritional, pomological and physical properties of cornelian cherry (Cornus mas L.). *Journal of Food Engineering*, *60*(3), 335–341.

Dhineshkumar, V., & Ramasamy, D. (2016). Pomegranate processing and value addition: A review. *Indian Horticulture Journal*, *6*(1), 1–12.

Dinda, B., Kyriakopoulos, A. M., Dinda, S., Zoumpourlis, V., Thomaidis, N. S., Velegraki, A., … & Dinda, M. (2016). Cornus mas L. (cornelian cherry), an important European and Asian traditional food and medicine: Ethnomedicine, phytochemistry and pharmacology for its commercial utilization in drug industry. *Journal of Ethnopharmacology*, *193*, 670–690.

Dolatkhahi, M., Dolatkhahi, A., & Nejad, J. B. (2014). Ethnobotanical study of medicinal plants used in Arjan–Parishan protected area in Fars Province of Iran. *Avicenna Journal of Phytomedicine*, *4*(6), 402.

Dueñas, M., Pérez-Alonso, J. J., Santos-Buelga, C., & Escribano-Bailón, T. (2008). Anthocyanin composition in fig (Ficus carica L.). *Journal of Food Composition and Analysis*, *21*(2), 107–115.

El-Rayes, D. A. (2009). Characterization of three date palm cultivars based on RAPD fingerprints and fruit chemical composition. *Meteorology, Environment and Arid Land Agriculture Sciences*, *20*(2), 3–20.

Espada-Bellido, E., Ferreiro-González, M., Carrera, C., Palma, M., Barroso, C. G., & Barbero, G. F. (2017). Optimization of the ultrasound-assisted extraction of anthocyanins and total phenolic compounds in mulberry (Morus nigra) pulp. *Food Chemistry*, *219*, 23–32.

Ezati, P., Khan, A., Rhim, J. W., Roy, S., & Hassan, Z. U. (2023). Saffron: Perspectives and sustainability for active and intelligent food packaging applications. *Food and Bioprocess Technology, 16*(6), 1177–1196.

Farahani, M., Salehi-Arjmand, H., Khadivi, A., & Akramian, M. (2019). Chemical characterization and antioxidant activities of Morus alba var. nigra fruits. *Scientia Horticulturae, 253*, 120–127.

Farhadi, K., Esmaeilzadeh, F., Hatami, M., Forough, M., & Molaie, R. (2016). Determination of phenolic compounds content and antioxidant activity in skin, pulp, seed, cane and leaf of five native grape cultivars in West Azerbaijan province, Iran. *Food Chemistry, 199*, 847–855.

Farmani, F., Moein, M., Khoshnood, M. J., & Sabahi, Z. (2018). Anthocyanin isolation from Berberis integerrima bunge fruits and determination of their antioxidant activity. *Free Radicals and Antioxidants, 8*(1), 1–5.

Fateh, M. V., Ahmed, S., Ali, M., & Bandyopadhyay, S. (2013). A review on the medicinal importance of pomegranate. *Journal of Pharmaceutical Sciences, 3*(4), 23–25.

Fekrat, H. (2003, October). The application of crocin and saffron ethanol-extractable components in formulation of health care and beauty care products. *Acta Horticulturae, 650*, 365–368.

Fonseca, L. R., Silva, G. R., Luís, Â., Cardoso, H. J., Correia, S., … & Socorro, S. (2021). Sweet cherries as anti-cancer agents: From bioactive compounds to function. *Molecules, 26*(10), 2941.

Fujii, S., Morita, Y., Ohta, T., Uto, T., & Shoyama, Y. (2022). Saffron (Crocus sativus L.) as a valuable spice and food product: A narrative review. *Longhua Chinese Medicine, 5*. https://doi.org/10.21037/lcm-22-1

Gazeran, S., Tajalli, F., & Mohamadi Sani, A. (2016). Effect of ultrasonic extraction on qualitative parameters of saffron edible extract. *Journal of Essential Oil Bearing Plants, 19*(5), 1286–1291.

Geow, C. H., Tan, M. C., Yeap, S. P., & Chin, N. L. (2021). A review on extraction techniques and its future applications in industry. *European Journal of Lipid Science and Technology, 123*(4), 2000302.

Ghafari, S., Fahimi, S., & Sahranavard, S. (2017). Plants used to treat hyperpigmentation in Iranian traditional medicine: A review. *Research Journal of Pharmacognosy, 4*(4), 71–85.

Ghafoor, K., Sarker, M. Z. I., Al-Juhaimi, F. Y., Babiker, E. E., Alkaltham, M. S., Almubarak, A. K., & Ahmed, I. A. M. (2022). Innovative and green extraction techniques for the optimal recovery of phytochemicals from Saudi date fruit flesh. *Processes, 10*(11), 2224.

Ghani, A., Amini, S., Mohtashami, S., & Neamati, S. H. (2022). Physicochemical and biochemical variations of jujube (Ziziphus jujuba Mill.) populations: A comparison of Iranian and imported types. *Industrial Crops and Products, 183*, 114898.

Ghosh, S., Sarkar, T., Das, A., & Chakraborty, R. (2022). Natural colorants from plant pigments and their encapsulation: An emerging window for the food industry. *LWT, 153*, 112527.

Gonçalves, A. C., Bento, C., Silva, B., Simões, M., & Silva, L. R. (2019). Nutrients, bioactive compounds and bioactivity: The health benefits of sweet cherries (Prunus avium L.). *Current Nutrition & Food Science, 15*(3), 208–227.

Habibi, A., Yazdani, N., Koushesh Saba, M., Chatrabnous, N., Molassiotis, A., Sarikhani, S., & Vahdati, K. (2023). Natural preservation and improving lipid oxidation inhibition of fresh walnut. *Horticulture, Environment, and Biotechnology, 64*(1), 133–142.

Hashmi, S. F., Saleem, H., Khurshid, U., Khursheed, A., Alam, M. T., Imran, M., ... & Ali Gill, M. S. (2024). Genus Berberis: A comprehensive and updated review on ethnobotanical uses, phytochemistry and pharmacological activities. *Chemistry & Biodiversity, 21*, e202400911.

Hassan, A., Shahbaz, H. M., Naeem, M., & Aadil, R. M. (2023). Food and non-food applications of berries' by-products. In Khalifa, I., & Nawaz, A. (Eds.), *Berry Bioactive Compound By-Products* (pp. 135–157). London: Academic Press.

Henry, B. S. (1996). Natural food colours. In Hendry, G. A. F., & Houghton, J. D. (Eds.), *Natural Food Colorants* (pp. 40–79). Boston, MA: Springer US.

Heydari, S., & Haghayegh, G. H. (2014). Extraction and microextraction techniques for the determination of compounds from saffron. *Canadian Chemical Transactions, 2*, 221–247.

Hosseini, S. H., Sadeghi, Z., Hosseini, S. V., & Bussmann, R. W. (2022). Ethnopharmacological study of medicinal plants in Sarvabad, Kurdistan province, Iran. *Journal of Ethnopharmacology, 288*, 114985.

Hosseinnezhad, M., & Safapour, S. (2024). Sources, chemistry, classification, challenges, and prospects of renewable dyes and pigments. In Ul Islam, S. (Ed.), *Renewable Dyes and Pigments* (pp. 1–18). Amsterdam: Elsevier.

Iqbal, M., Panhwar, A., Ahmed, K., Kandhro, A., Sultana, R., Mughal, J., & Solangi, Z. (2022). Textile dyeing of cotton and wool textile material with natural dyes extracted from bluish purple grapes. *Bulgarian Chemical Communications, 54*(1), 14–18.

Jafarpour, M., Yousefi, G., Hamedi, A., Shariat, A., Salehi, A., & Heydari, M. (2016). Effect of a traditional syrup from Citrus medica L. fruit juice on migraine headache: A randomized double-blind placebo controlled clinical trial. *Journal of Ethnopharmacology, 179*, 170–176.

Kandylis, P., & Kokkinomagoulos, E. (2020). Food applications and potential health benefits of pomegranate and its derivatives. *Foods, 9*(2), 122.

Karjalainen, R., Anttonen, M., Saviranta, N., Stewart, D., McDougall, G. J., Hilz, H., ... & Törrönen, R. (2008). A review on bioactive compounds in black currants (Ribes nigrum L.) and their potential health-promoting properties. *Acta Horticulturae, 839*, 301–307.

Kasapoğlu, E. D., Kahraman, S., & Törnük, F. (2020). Apricot juice processing byproducts as sources of value-added compounds for food industry. *European Food Science and Engineering, 1*(1), 18–23.

Khoo, H. E., Azlan, A., Tang, S. T., & Lim, S. M. (2017). Anthocyanidins and anthocyanins: Colored pigments as food, pharmaceutical ingredients, and the potential health benefits. *Food & Nutrition Research, 61*, 1361779.

Kiumarsi, A., & Gashti, M. P. (2015). Pistachio hulls, a new source of fruit waste for wool dyeing. *Journal of Textile Science & Engineering, 5*(1), 1.

Kothari, D., Thakur, R., & Kumar, R. (2021). Saffron (Crocus sativus L.): Gold of the spices—A comprehensive review. *Horticulture, Environment, and Biotechnology, 62*(5), 661–677.

Kucelova, L., Grygorieva, O., Ivanišová, E., Margarita, T., & Brindza, J. (2016). Biological properties of black mulberry-derived food products (Morus nigra L.). *Journal of Berry Research*, *6*(3), 333–343.

Kulkarni, S. S., Gokhale, A. V., Bodake, U. M., & Pathade, G. R. (2011). Cotton dyeing with natural dye extracted from pomegranate (Punica granatum) peel. *Universal Journal of Environmental Research & Technology*, *1*(2), 135–139.

Leal, C., Gouvinhas, I., Santos, R. A., Rosa, E., Silva, A. M., Saavedra, M. J., & Barros, A. I. (2020). Potential application of grape (Vitis vinifera L.) stem extracts in the cosmetic and pharmaceutical industries: Valorization of a by-product. *Industrial Crops and Products*, *154*, 112675.

Lemoine, M. L., Fontana, U., Hurtado, J. B., Pintos, F. M., Arena, M. E., Vicente, A. R., & Rodoni, L. M. (2021). Wild barberry fruit (berberis microphylla g. Forst.) as a natural ingredient for beer brewing. *Chilean Journal of Agricultural & Animal Sciences*, *37*(3), 313–324.

Lorenz, P., Berger, M., Bertrams, J., Wende, K., Wenzel, K., Lindequist, U., … & Stintzing, F. C. (2008). Natural wax constituents of a supercritical fluid CO_2 extract from quince (Cydonia oblonga Mill.) pomace. *Analytical and Bioanalytical Chemistry*, *391*, 633–646.

Lotfi, L., Kalbasi-Ashtari, A., Hamedi, M., & Ghorbani, F. (2015). Effects of enzymatic extraction on anthocyanins yield of saffron tepals (Crocos sativus) along with its color properties and structural stability. *Journal of Food and Drug Analysis*, *23*(2), 210–218.

Mahdavi, S. A., Jafari, S. M., Assadpour, E., & Ghorbani, M. (2016). Storage stability of encapsulated barberry's anthocyanin and its application in jelly formulation. *Journal of food engineering*, *181*, 59–66.

Mahesh, S. K., Fathima, J., & Veena, V. G. (2019). Cosmetic potential of natural products: Industrial applications. In Swamy, M. K., & Akhtar, M. S. (Eds.), *Natural Bio-Active Compounds: Volume 2: Chemistry, Pharmacology and Health Care Practices* (pp. 215–250). Singapore: Springer.

Mahmoodi, S., Moradinezhad, F., Moodi, S., Amini, H., & Khayyat, M. (2018). Effects of potassium metabisulfite pre-treatment and different drying temperatures on some chemical properties and color retention of Russian olive (Elaeagnus angustifolia) fruit. *Journal of Horticulture and Postharvest Research*, *1*, 49–62.

Maier, T., Göppert, A., Kammerer, D. R., Schieber, A., & Carle, R. (2008). Optimization of a process for enzyme-assisted pigment extraction from grape (Vitis vinifera L.) pomace. *European Food Research and Technology*, *227*, 267–275.

McCune, L. M., Kubota, C., Stendell-Hollis, N. R., & Thomson, C. A. (2010). Cherries and health: A review. *Critical Reviews in Food Science and Nutrition*, *51*(1), 1–12.

Milea, A. Ș., Vasile, A. M., Cîrciumaru, A., Dumitrașcu, L., Barbu, V., Râpeanu, G., … & Stănciuc, N. (2019). Valorizations of sweet cherries skins phytochemicals by extraction, microencapsulation and development of value-added food products. *Foods*, *8*(6), 188.

Mirshahi, F., Khosravi, A., Gharanjig, K., & Fakhari, J. (2013). Antimicrobial properties of treated cotton fabrics with non-toxic biopolymers and their dyeing with safflower and walnut hulls. *Iranian Polymer Journal*, *22*, 843–851.

Mobli, M., Haririan, I., Amin, G., Kamalinejad, M., & Hajimahmoodi, M. (2016). Stability of polyphenols in myrtle berries syrup, a traditional Iranian medicine, a Traditional Iranian Medicine. *Traditional and Integrative Medicine, 1*(1), 35–39.

Mollaamin, F., Mohammadian, N. T., Najaflou, N., & Monajjemi, M. (2021). Iranian Qara Qat fruit (redcurrant) in Arasbaran forests as the resource of anthocyanin pigments in formation of [ACN-Mg^{2+}/Al^{3+}/Ga^{3+}/Sn^{2+}/Cr^{3+}/Fe^{3+}] chelation clusters. *SN Applied Sciences, 3*, 1–18.

Moshiri, M., Vahabzadeh, M., & Hosseinzadeh, H. (2015). Clinical applications of saffron (Crocus sativus) and its constituents: A review. *Drug Research, 65*(06), 287–295.

Mostafavi, H., Pezhhanfar, S., Fouladi, F., & Agayi, F. (2021). Investigation of the phytochemical constituents and antimicrobial potential of the fruit and flower of Pistacia atlantica Desf. as a herbal medicine growing in north-west of Iran. *Journal of Medicinal Plants and By-product, 10*(2), 237–245.

Motaghi, Z. (2018). An economical dyeing process for cotton and wool fabrics and improvement their antibacterial properties and UV protection. *Journal of Natural Fibers, 15*(6), 777–788.

Naji-Tabasi, S., Emadzadeh, B., Shahidi-Noghabi, M., Abbaspour, M., & Akbari, E. (2021). Physico-chemical and antioxidant properties of barberry juice powder and its effervescent tablets. *Chemical and Biological Technologies in Agriculture, 8*, 1–11.

Najman, K., Adrian, S., Sadowska, A., Świąder, K., Hallmann, E., Buczak, K., … & Szterk, A. (2023). Changes in physicochemical and bioactive properties of quince (Cydonia oblonga Mill.) and its products. *Molecules, 28*(7), 3066.

Nawirska-Olszańska, A., Biesiada, A., Sokół-Łętowska, A., & Kucharska, A. Z. (2011). Content of bioactive compounds and antioxidant capacity of pumpkin puree enriched with Japanese quince, cornelian cherry, strawberry and apples. *Acta Scientiarum Polonorum Technologia Alimentaria, 10*(1), 51–60.

Nobari, A., Marvizadeh, M. M., Sadeghi, T., Rezaei-Savadkouhi, N., & Mohammadi Nafchi, A. (2022). Flavonoid and anthocyanin pigments characterization of pistachio nut (Pistacia vera) as a function of cultivar. *Journal of Nuts, 13*(4), 313–322.

Pasdaran, A., Zare, M., Hamedi, A., & Hamedi, A. (2023). A review of the chemistry and biological activities of natural colorants, dyes, and pigments: Challenges, and opportunities for food, cosmetics, and pharmaceutical application. *Chemistry & Biodiversity, 20*(8), e202300561.

Patel, S. (2014). Plant genus Elaeagnus: Underutilized lycopene and linoleic acid reserve with permaculture potential. *Fruits, 70*(4), 191–199.

Petrov Ivanković, A., Milivojević, A., Ćorović, M., Simović, M., Banjanac, K., Jansen, P., … & Bezbradica, D. (2023). In vitro evaluation of enzymatically derived blackcurrant extract as prebiotic cosmetic ingredient: Extraction conditions optimization and effect on cutaneous microbiota representatives. *Chemical and Biological Technologies in Agriculture, 10*(1), 125.

Popescu, V. (2024). New trends in the application of natural dyes in textile dyeing. In Ul-Islam, S. (Ed.), *Renewable Dyes and Pigments* (pp. 111–137). Amsterdam: Elsevier.

Rymbai, H., Sharma, R. R., & Srivastav, M. (2011). Bio-colorants and its implications in health and food industry–A review. *International Journal of Pharmacological Research, 3*(4), 2228–2244.

Sadeghi, Z., Valizadeh, J., & Shermeh, O. A. (2015). Antioxidant activity and total phenolic contents of some date varieties from Saravan Region, Baluchistan, Iran. *Journal of Medicinal Plants Research*, *9*(4), 78–83.

Sadeghi-Kiakhani, M., Tehrani-Bagha, A. R., Safapour, S., Eshaghloo-Galugahi, S., & Etezad, S. M. (2021). Ultrasound-assisted extraction of natural dyes from Hawthorn fruits for dyeing polyamide fabric and study its fastness, antimicrobial, and antioxidant properties. *Environment, Development and Sustainability*, *23*, 9163–9180.

Sadighara, P., Godarzi, S., Bahmani, M., & Asadi-Samani, M. (2016). Antioxidant activity and properties of walnut brown seed coat extract. *Journal of Global Pharma Technology*, *8*(11), 26–30.

Seeram, N. P., Schutzki, R., Chandra, A., & Nair, M. G. (2002). Characterization, quantification, and bioactivities of anthocyanins in Cornus species. *Journal of Agricultural and Food Chemistry*, *50*(9), 2519–2523.

Shakeri, A., Soheili, V., Karimi, M., Hosseininia, S. A., & Fazly Bazzaz, B. S. (2018). Biological activities of three natural plant pigments and their health benefits. *Journal of Food Measurement and Characterization*, *12*, 356–361.

Sharayei, P., Azarpazhooh, E., Zomorodi, S., & Ramaswamy, H. S. (2019). Ultrasound assisted extraction of bioactive compounds from pomegranate (Punica granatum L.) peel. *LWT*, *101*, 342–350.

Sharifi, A., & Hassani, B. (2012). Extraction methods and stability of color extracted from barberry pigments. *International Journal of AgriScience*, 2(4), 320–327.

Sheng, F., Wang, Y., Zhao, X., Tian, N., Hu, H., & Li, P. (2014). Separation and identification of anthocyanin extracted from mulberry fruit and the pigment binding properties toward human serum albumin. *Journal of Agricultural and Food Chemistry*, *62*(28), 6813–6819.

Soltani, R., Gorji, A., Asgary, S., Sarrafzadegan, N., & Siavash, M. (2015). Evaluation of the effects of Cornus mas L. fruit extract on glycemic control and insulin level in type 2 diabetic adult patients: A randomized double-blind placebo-controlled clinical trial. *Evidence-Based Complementary and Alternative Medicine*, *2015*(1), 740954.

Stabnikova, O., Stabnikov, V., & Paredes-López, O. (2024). Fruits of wild-grown shrubs for health nutrition. *Plant Foods for Human Nutrition*, *79*(1), 20–37.

Sun, Y., Li, S., Zeng, F., Qi, J., Qin, W., Tan, C., … & Chen, H. (2019). Functional components, antioxidant activity and hypoglycemic ability following simulated gastro- intestinal digestion of pigments from walnut brown shell and green husk. *Antioxidants*, *8*(12), 573.

Tehranifar, A., Zarei, M., Esfandiyari, B., & Nemati, Z. (2010). Physicochemical properties and antioxidant activities of pomegranate fruit (Punica granatum) of different cultivars grown in Iran. *Horticulture Environment and Biotechnology*, *51*(6), 573–579.

Thamrongwatwongsa, J., Pattarapipatkul, N., Jaithon, T., Jindaruk, A., Paemanee, A., T-Thienprasert, N. P., & Phonphoem, W. P. (2022). Mulberroside F from in vitro culture of mulberry and the potential use of the root extracts in cosmeceutical applications. *Plants*, *12*(1), 146.

Venkitasamy, C., Zhao, L., Zhang, R., & Pan, Z. (2019). Pomegranate. In Pan, Z., Zhang, R., & Zicari, S. (Eds.), *Integrated Processing Technologies for Food and Agricultural By-Products* (pp. 181–216). London: Academic Press.

Wang, H., Hussain, M., & Zhou, W. (2017). A novel method for natural dyeing of cotton fabrics with anthocyanin pigments from Morus rubra fruits. *Textile Research Journal, 87*(15), 1896–1911.

Wojdyło, A., Oszmiański, J., Teleszko, M., & Sokół-Łętowska, A. (2013). Composition and quantification of major polyphenolic compounds, antioxidant activity and colour properties of quince and mixed quince jams. *International Journal of Food Sciences and Nutrition, 64*(6), 749–756.

Xue, B., Hui, X., Chen, X., Luo, S., Dilrukshi, H. N. N., Wu, G., & Chen, C. (2022). Application, emerging health benefits, and dosage effects of blackcurrant food formats. *Journal of Functional Foods, 95*, 105147.

Yasukawa, A., Chida, A., Kato, Y., & Kasai, M. (2017). Dyeing silk and cotton fabrics using natural blackcurrants. *Textile Research Journal, 87*(19), 2379–2387.

Zahed, N., Esmaeilzadeh Kenari, R., & Farahmandfar, R. (2023). Effect of different extraction methods on antioxidant properties and encapsulation efficiency of anthocyanin of pomegranate peel. *Food Science & Nutrition, 11*(7), 3780–3787.

Zam, W., Ali, A., & Ibrahim, W. (2017). Improvement of polyphenolic content and antioxidant activity of Syrian myrtle berries (Myrtus communis L.) hydro-alcoholic extracts using flavoring additives. *Diabetes, 26*, 27.

Zare, H., & Sotoudeh, M. M. (2018). Effects of harvesting stages and storage temperatures on pigment contents in fresh fig of four Iranian cultivars. *Journal of Chemical Health Risks, 8*(2), 147–156.

Zhou, X. J., Zhu, C. T., Zhang, L. Y., You, S., Wu, F. A., & Wang, J. (2021). Enrichment and purification of red pigments from defective mulberry fruits using biotransformation in a liquid-liquid-solid three-phase system. *Environmental Science and Pollution Research, 28*, 24432–24440.

chapter four

Extraction of antioxidants from medicinal fruits of Iraq

Mehvish Habib, Kulsum Jan, Eram S. Rao, and Khalid Bashir

4.1 Introduction

People from all over the world, especially developing nations like Iraq, have been relying on herbal medicine and traditional phytotherapies for generations as a means of treating a broad range of illnesses and disorders, including cancer, diabetes, malaria, flu, intestinal problems, blood pressure, kidney, and heart diseases (Pourmorad et al., 2006). It's intriguing to observe that there is a global movement towards a more natural lifestyle these days and that traditional phytotherapies are valued for their safety as well as the range of adverse effects associated with the synthetic medications that are currently on the market. The growing popularity of herbal medicine, as well as the growth of herbal shops and health food stores, is indicative of the growing interest in complementary and alternative medicines (Berman et al., 1999).

It is commonly recognized that a variety of syndromes and diseases, including diabetes, atherosclerosis, cancer, and liver cirrhosis, are caused by free radicals. Additionally, free-radical scavenging agents have the capability to shield the human body from the harm caused by free radicals by reducing the detrimental effects of these conditions (Wilson, 1998). Diabetes patients have been found to experience oxidative stress, which is accompanied by a decline in antioxidant status. This combination can exacerbate the harm that free radicals can do to bodily tissues and cells (Collier et al., 1990). The most dangerous chronic metabolic disease, diabetes mellitus, is becoming more and more commonplace and is a global health concern. About 95% of cases of diabetes mellitus are type 2, making it the most common type (Cheplick et al., 2010). There are currently more than 170 million diabetics worldwide, and by 2030, that number is predicted to double (Boon et al., 2006). Diabetes mellitus is linked to a number of complications, including atherosclerosis, myocardial infarction, neuropathy, and nephropathy. These complications seem to be correlated

 DOI: 10.1201/9781003452614-4

with the level of hyperglycemia and the oxidative stress that follows (Maritim et al., 2003). Furthermore, the majority of artificial substances used to cause diabetes, like alloxan, work by causing oxygen free radicals to be produced, which harm the pancreas (Sabu et al., 2002). However, it has been demonstrated that antioxidants lower oxidative stress in experimental diabetes, which may help to lessen the complications related to the disease (Garg et al., 2000).

Alternative safe and effective therapeutic approaches that have few or no side effects are desperately needed, as the currently prescribed treatments have a broad variety of negative effects. Medicinal plants have long been utilized to treat postprandial hyperglycemia by means of their phytochemicals or leaves (Campbell-Hunt, 2000). It is widely recognized that one of the most frequently identified signs of type 2 diabetes is postprandial hyperglycemia (Bell, 2001). This condition arises when the pancreas is unable to produce enough insulin (Chiasson et al., 2002). It is commonly known that the main source of blood glucose is starch, and that the amount and digestion of starch eaten are related factors in postprandial hyperglycemia. As a result, postprandial hyperglycemia can be managed by inhibiting the enzymes that hydrolyze starch, which reduces or slows digestion along with the absorption of dietary carbohydrates (McDougall et al., 2005). In the small intestine, α-amylase breaks down starch to produce oligosaccharides, which are then further hydrolyzed into glucose by α-glucosidases, another enzyme (Casirola and Ferraris, 2006). Therefore, compounds derived from medicinal plants and/or their extracts that possess the capability to inhibit the α-glucosidase enzyme could be promising options for the treatment of diabetes.

Because there are so many native fruits in Iraq that have the potential to have health benefits, it is important to extract antioxidants from these fruits in order to make medicinal products. To extract and concentrate antioxidants from the many fruits that grow in the area, including pomegranates, dates, and figs, different extraction techniques are used in this process. These antioxidants, which include vitamins, flavonoids, and polyphenols, have exceptional abilities to scavenge free radicals, which may help explain some of their possible medical benefits. To obtain high-quality antioxidant extracts, researchers frequently use techniques like solvent extraction, supercritical fluid extraction, or microwave-assisted extraction.

4.1.1 Foeniculum vulgare

Foeniculum vulgare Mill. is a biennial medicinal and aromatic plant belonging to the family Apiaceae (*Umbelliferaceae*). It is a resilient, evergreen herb with feathery leaves and yellow flowers. Its hollow stems enable it to reach a maximum height of 2.5 m. The leaves can reach a length of 40 cm and are finely divided, with the final segments being filiform, or thread-like, and measuring roughly 0.5 mm in width. The final compound umbels from which the flowers are produced. The fruit is a 4–10 mm long, dry seed. Though it is primarily found on dry soils close to the coast as well as on riverbanks, it is generally thought to be native to the Mediterranean Sea's shores. However, it has spread throughout the world. According to Díaz-Maroto et al. (2006), there are two subspecies of fennel, *vulgare and piperitum*. The former has bitter seeds, while the latter has sweet seeds, which have been utilized as flavoring agents in baked goods, fish and meat dishes, ice creams, alcoholic beverages, and other products because of their distinct anise odor. The distinctions in morphology between these two subspecies are not always obvious.

4.1.2 Culinary uses

This herb has both culinary and medicinal applications. It has a strong aroma and flavor. With an aroma akin to anise, fennel seeds have been utilized as flavorings in ice cream, meat, and fish dishes, baked goods, mixtures of herbs, and alcoholic beverages (Diaaz-Maroto et al., 2005). Many international culinary traditions make extensive use of the fennel plant's bulb, foliage, and seeds. Fresh fennel seed is brown or green in color, but as it ages, it gradually turns a dull grey color. Dried fennel seed has an aromatic, anise-flavored taste. Green seeds work best when cooked. Crisp and resilient, the bulb can be prepared in a variety of ways: sautéed, braised, stewed, grilled, or eaten raw. Mediterranean cuisine uses fennel primarily in salad dressings, pasta dishes, vegetable dishes, and side dishes. Both the cooked and raw fronds and bulbs are used in these recipes. Fennel seeds are used in cooking by many Middle Eastern and Indian subcontinental cultures. The most vital spices in Gujarati and Pandit cuisine from Kashmir are fennel (Grieve, 1931).

4.1.3 Use in local and traditional medicine

Due to its carminative properties, fennel is mainly employed for medicinal purposes to alleviate the adverse effects of purgatives. It is a component of the well-known blend known as liquorice powder. Fennel water exhibits similar traits to both anise as well as dill water. When syrup and sodium bicarbonate are mixed together, they create a common household remedy called "gripe water," which is used to alleviate flatulence in infants. Fennel tea, which also acts as a carminative, has been made by pouring boiling water over a crushed teaspoon of fennel seeds. Fennel seeds are ingested in their raw form on the Indian Subcontinent and sometimes sweetened to improve eyesight. Fennel seed extracts have shown promise as a potential treatment for glaucoma, hypertension, and as a diuretic in animal experiments. It has been utilized as a galactagogue to enhance a lactating mother's lactation. The existence of phytoestrogens in fennel may contribute to the stimulation of breast tissue growth (Agarwal et al., 2008).

4.1.4 *Phytochemistry*

F. vulgare has been reported to contain 9.5% protein, 6.3% moisture, 13.4% minerals, 10% fat, 42.3% carbohydrates, and 18.5% fiber. *F. vulgare* contains a variety of minerals and vitamins, including potassium, calcium, sodium, iron, thiamine, phosphorus, riboflavin, niacin, and vitamin C.

4.1.5 *Essential oil*

For its essential oil, *F. vulgare* is widely recognized. Because of its essential oil, *F. vulgare* has a distinctive anise scent that makes it a great flavoring for baked goods, fish and meat dishes, alcoholic beverages, and ice cream. The major components of *F. vulgare* seed essential oil have been reported to be trans-anethole, fenchone, estragol (methyl chavicol), and α-phellandrene; their molecular structures are shown in Figure 4.1. The relative concentration of these compounds varies considerably depending

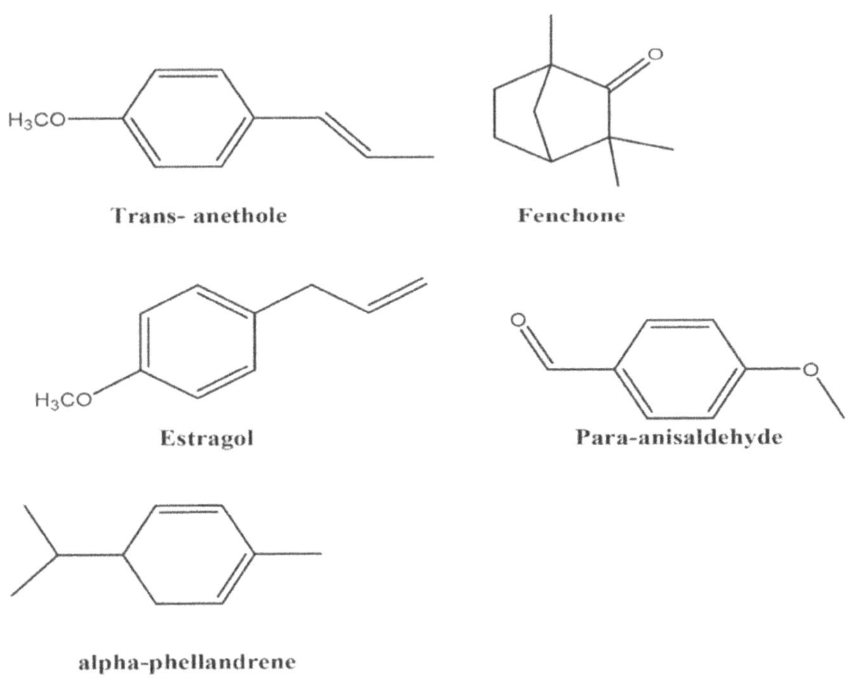

Figure 4.1 Molecular structures of the major bioactive essential oil components of *F. Vulgare*.

on the phonological state and origin of the fennel (Diaaz-Maroto et al., 2006). Depending on the extraction process and the plant's origin, the *F. vulgare* essential oil composition varies greatly. These volatile substances build up inside the plant in a variety of places, almost anywhere—in the roots, shoots, stem, flowers, and fruits, for example (Diaaz-Maroto et al., 2006; Gross et al., 2009). According to one study, the *F. vulgare* essential oil composition and content change as the plant reaches different stages of maturity. With fruit maturity, there has been a reported decline in the essential oil content. The primary component, trans-anethole, had a content ranging from 81.63% to 87.85% (Telci et al., 2009). The primary oleoresin constituents of the aerial parts of *F. vulgare, estragol and trans-anethole,* were found to vary during plant development, with the concentration of these two compounds being highest in flowers along with the growing mericarps, according to a different study. The essential oil found in *F. vulgare*'s fruits is typically responsible for the pharmacological effects of these fruits. Several investigations have demonstrated the essential oil's unique pharmacological properties as well as those of each of its constituent parts. The two main acaricidal agents against *Dermatophagoides farinae* and *Dermatophagoides pteronyssinus* have been observed to be *fentanyl* and *panisaldehyde.* Therefore, these compounds have the potential to be used as lead compounds or as agents for controlling house dust mites. *Anethole* has been reported to be an active oestrogenic agent in another study. Nonetheless, certain research has demonstrated that the true oestrogenic agents are anethole polymers, such as *dianethole and photoanethole.* Because of its clot-destabilizing effect, antiplatelet activity, and vasorelaxant action, anethole has also been reported to be a safe antithrombotic agent (Tognolini et al., 2007). But since *methyleugenol,* which shares a structural similarity with estragole, was recently discovered to be potentially carcinogenic, estragole—*F. vulgare*'s major component—became a source of concern. As a result, a new legal limit of 10 mg/kg for estragole in non-alcoholic beverages has been approved by the European Union (EU) (Zeller and Rychlik, 2006). The chemical structures of some of the phytochemicals and other classes found in *F. vulgare,* including phenolic glycosides and phenols, are displayed in Figure 4.2.

Figures 4.3 and 4.4 show the structures of a few flavonoid aglycones and other compounds. Phenolic acids such as 3-*O*-Caffeoylquinic acid, 4-*O*-Caffeoylquinic acid, 5-*O*-Caffeoylquinic acid, 1,3-*O*-di-Caffeoylquinic acid, 1,4-*O*-di-Caffeoylquinic acid, and 1,5-*O*-di-Caffeoylquinic acid have been reported to be present in *F. vulgare.* From *F. vulgare,* flavonoids such as rosmarinic acid, quercetin-3-rutinoside, and eriodictyol-7-rutinoside have also been isolated (Faudale et al., 2008; Park, 1996). There have also been reports of kaempferol-3-*O*-rutinoside, quercetin-3-*O*-galactoside, and kaempferol-3-*O*-glucoside occurring in the *F. vulgare* aqueous extract. Further isolated compounds from *F. vulgare*

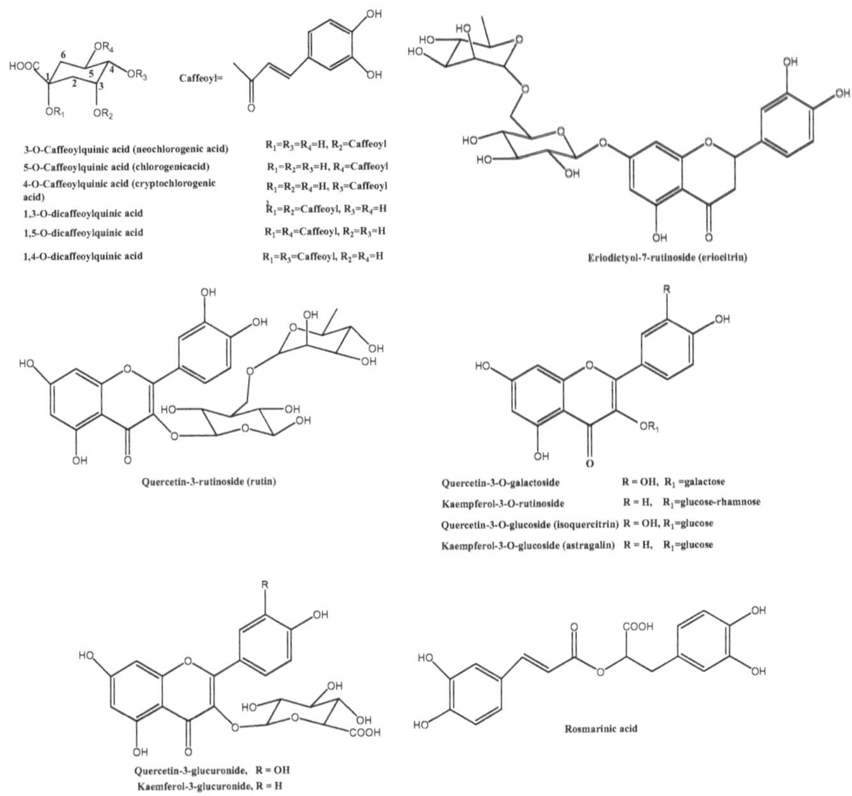

Figure 4.2 Molecular structure of some phenols and phenolic glycosides isolated from *F. Vulgare.*

include kampferol-3-*O*-glucuronide, quercitin-3-*O*-glucuronide, isoquercitin, and isorhamnetin-3-*O*-glucoside (Parejo et al., 2004a). The phenolic compounds found in *F. vulgare* are thought to have a role in preventing conditions like cancer, inflammation, and cardiovascular disease that are believed to be brought on by oxidative stress. Because of these phenolic compounds' roles in human health, nutritionists, food scientists, and consumers have given them a lot of attention. Along with trans-re sveratrol-3-*O*-β-D-glucopyranoside, cis-miyabenol C, trans-miyabenol C, sinapyl glucoside, syringin-4-*O*-β-glucoside, oleanolic acid, 5,8-epidioxy-ergosta-6,22-dien-3-ol, 7a-hydroxycampesterol, (3β,5α,8α,22E)", and 2,3-dihydropropylheptadec-5-onoate, had also been isolated from *F. vulgare* fruit (Marino et al., 2007).

F. vulgare flowers yield an acylated kaemferol glycoside (Soliman et al., 2002) (Figures 4.3 and 4.5).

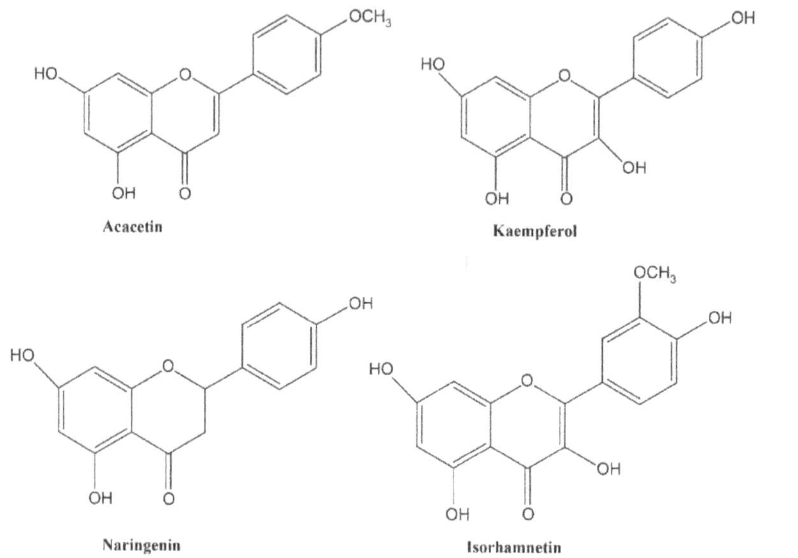

Figure 4.3 Molecular structures of some flavonoid aglycons reported in *F. Vulgare.*

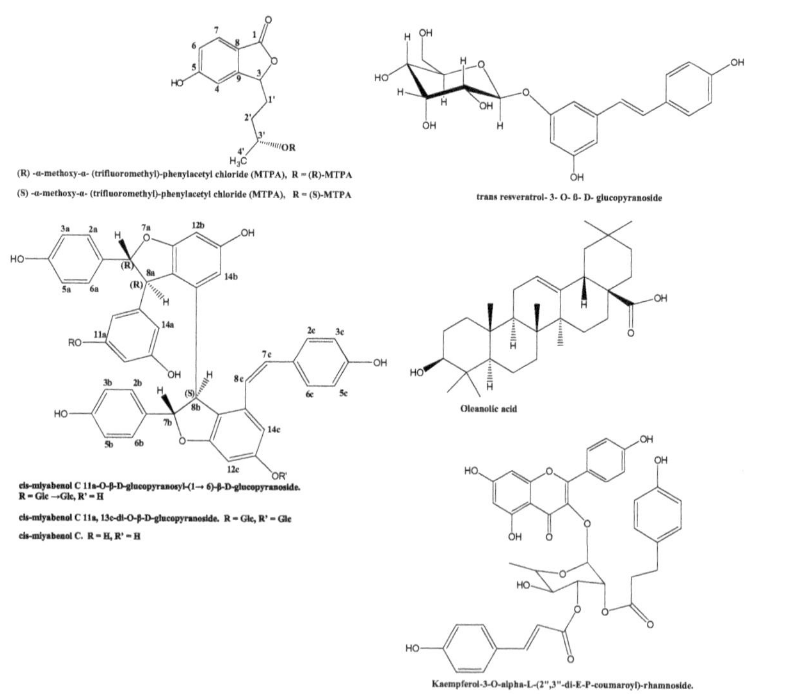

Figure 4.4 Molecular structures of some miscellaneous compounds isolated from *F. Vulgare.*

4.2 *Pharmacology*

4.2.1 *Antibacterial activity*

Foodborne pathogens like *E. coli, B. megaterium, and S. aureus* (Mohsenzadeh, 2007), *E. coli 0157:H7, S. aureus, and Listeria monocytogenes* (Dadalioglu and Evrendilek, 2004; Cantore et al., 2004) were all susceptible to the antibacterial effect of the essential oil that has been extracted from the fruits of *F. vulgare*. It has been observed that *F. vulgare* exhibits antibacterial activity against certain bacterial strains in both aqueous and organic extracts (Kaur and Arora, 2008). There have also been reports of *F. vulgare* seed essential oil's antibacterial activity against certain human pathogenic bacteria. *F. vulgare* ethanol and water extracts have demonstrated efficacy against *Campylobacter jejuni* and *Helicobacter pylori* (Mahady et al., 2005). The essential oil of *F. vulgare* has demonstrated potential in managing *Acinetobacter baumannii* infections that are resistant to multiple drugs in a different study. A phenylpropanoid derivative called *Dillapional* was discovered to be the *F. vulgare* stem's active antimicrobial principle, among other chemical constituents from the plant that had been identified as active antimicrobial principles. Another molecule isolated from *F. vulgare* is called *scopoletin*, a coumarin derivative that has been shown to have a weak antibacterial effect (Kwon et al., 2002).

4.2.2 *Antifungal activity*

It has been observed that fennel essential oil has antifungal properties. There have been reports of antimycobacterial and anticandidal properties in fennel essential oil along with the seed extracts (Abed, 2007). There have also been reports of *F. vulgare* bark extracts having antifungal activity against *C. albicans* (Pai et al., 2010). According to reports, *F. vulgare* essential oil inhibits *Sclerotinia sclerotium*'s mycelial growth and germination. As a result, it may be utilized as a biofungicide substitute for synthetic fungicides when controlling phytopathogenic fungi (Soylu et al., 2007). According to Singh et al. (2006), *F. vulgare's* essential oil displays a complete ZOI against *Aspergillus flavus, Aspergillum niger, Fusarium graminearum*, and *F. moniliforme* at a dosage of 6 μL.

4.2.3 *Antioxidant activity*

It has been established how much antioxidant power is present in edible, wild, and medicinal fennel plants from various Mediterranean nations. It has been discovered that the radical scavenging activity of wild fennel is greater than that of edible as well as medicinal fennel (Faudale et al., 2008). Additionally, it has been observed that the *F. vulgare* fruit methanolic extract exhibits antioxidant activity by lowering the level of malondialdehyde in the *F. vulgare* fruit methanol extract group compared to the control group. According to reports, *F. vulgare's* essential oil along with the extracts of acetone have significant antioxidant activity when compared to butylated hydroxytoluene (BHT) and butylated hydroxyanisole (BHA) (Ruberto et al., 2000). By using the ferric thiocyanate method to monitor peroxide accumulation in the emulsion at the time of incubation, the inhibitory effects of oil along with the acetone extracts in the linoleic acid system were investigated. Fruit extract from *F. vulgare* and its purified constituents, cis-miyabenol C 11a-*O*-β-D-Antioxidant activity has been reported for -glucopyranosyl-(1fi6)-β-D-glucopyranoside, sinapyl glucoside, cis- and trans-miyabenol C, and syringin-4-*O*-β-glucoside, have been reported to show antioxidant activity. In the lipid peroxidation test, the *F. vulgare* fruit *n*-BuOH extract exhibited strong activity at the greater tested concentration but only moderate activity otherwise. Compared to crude extracts, pure compounds isolated from *F. vulgare* demonstrated greater antioxidant activity (Marino et al., 2007).

The residue of bitter fennel's flowering aerial parts, left after distillation for essential oils, contains isolated phenolic compounds that have demonstrated potent antiradical scavenging activity. These compounds can potentially elucidate the pharmacological impact of *F. vulgare.* The isolated compounds, which have been identified as rosmarinic acid, eriodictyol-7-rutinoside, kaempferol-3-*O*-rutinoside, quercetin-3-*O*-galactoside, 1,5-*O*-dicaffeoylquinic acid, 4-caffeoylquinic acid, and kaempferol-3-*O*-glucoside, were reported by Parejo et al., 2004b. The antioxidant activity of *F. vulgare* seed extracts in ethanol as well as water extracts has also been reported. Within the linoleic acid system, the water and ethanol extracts exhibited 99.1% and 77.5% inhibition of peroxidation, respectively, when 100 μg of each was used. These inhibition rates were higher than that of α-tocopherol, which showed a peroxidation inhibition rate of 36.1% at the same dose. Both extracts exhibited significant efficacy in decreasing power, superoxide anion radicals, scavenging free radicals, hydrogen peroxide, and chelating metal ions. According to reports, the fruits of three Egyptian fennel cultivars, namely *F. vulgare* var. dulce, *F. vulgare* var. azoricum, and *F. vulgare* var. vulgare, grown using organic methods, possess antioxidant properties in their essential oils. Shahat et al. (2011) discovered that the essential oils derived from the dulce and azoricum cultivars exhibited greater antioxidant activity compared to those obtained from the vulgare cultivar.

4.2.4 *Antithrombotic activity*

The demonstrated safe antithrombotic activity of *F. vulgare* essential oil, along with its primary constituent, anethole, stems from their broad-spectrum antiplatelet activity, vasorelaxant action, and clot destabilizing effect. In guinea pig plasma, anethole, the primary constituent of fennel oil, demonstrated the same level of potency as fennel oil in terms of inhibiting aggregation induced by U46619, collagen-ADP, and arachidonic acid. At concentrations comparable to those of fennel oil, anethole also inhibited the thrombin-induced clot reaction. In rat aortas with or without endothelium, fennel oil and anethole demonstrated similar NO-independent vasorelaxant activity at antiplatelet concentrations that have been shown to be free of cytotoxic effects in vitro. Moreover, rats' ethanol-induced stomach lesions were significantly prevented by both the *F. vulgare* essential oil and anethole (100 mg/kg oral administration) (Tognolini et al., 2007).

4.2.5 *Anti-inflammatory activity*

It has been observed that *F. vulgare* fruit methanolic extract (200 mg/kg) oral administration exhibits inhibitory effects against type-IV allergic reactions as well as acute and subacute inflammatory diseases (Choi and Hwang, 2004).

4.2.6 *Oestrogenic activity*

For centuries, *F. vulgare* has been utilized as an oestrogenic agent. It has been claimed to boost libido, ease and encourage menstruation symptoms, alleviate male climacteric, enhance milk secretion, and facilitate childbirth. Anethole, the fennel essential oil's primary component, is thought to be the active oestrogenic agent. According to some other research, anethole polymers like dianethole and photoanethole are the real pharmacologically active substances (Albert-Puleo, 1980).

4.2.7 *Hepatoprotective activity*

It has been observed that fennel essential oil has hepatoprotective properties. Fennel essential oil was shown in a study to reduce alanine aminotransferase (ALT), alkaline phosphatase (ALP), serum levels of bilirubin, and aspartate aminotransferase (AST), thereby attenuating the hepatotoxicity caused by the acute CCl_4 administration (Ozbek et al., 2003).

4.2.8 *Miscellaneous*

4.2.8.1 *In "vitro cytoprotection and antitumor activity*

Reports indicate that the *F. vulgare* methanolic extract demonstrates antitumor activity against the B16F10 melanoma cell line, as determined by the trypan blue exclusion test for cell viability. Additionally, the extract exhibits in vitro cytoprotective activity against normal human blood lymphocytes, as assessed by the micronucleus assay. When treated with 70% methanolic extract of F. vulgare, lymphocyte cultures displayed a micronucleus percentage of 0.006%, which is significantly lower than the 0.018% observed when treated with the standard drug doxorubicin. Conversely, at a concentration of 200 µg/mL, a 70% *F. vulgare* methanolic extract exhibited good anti-tumor activity. According to Pradhan et al. (2008), this implies that *F. vulgare* may be regarded as a natural source of antitumor agents in addition to being cytoprotective to normal cells.

4.2.8.2 *Acaricidal activity*

It has been reported that *F. vulgare* fruit oil exhibits acaricidal activity when applied directly to *D. pteronyssinus* and *D. farinae.* This activity has been compared to that of the commercial repellent benzyl benzoate. (+)-fenchone, *P*-anisaldehyde, (–)-fenchone, estragol, and thymol have been identified as the biologically active components of the *F. vulgare* fruit oil (Lee, 2004).

Using skin and patch tests, it has been observed that the *F. vulgare* fruit's methanol extract exhibits mosquito-repellent activity against female *Aedes aegypti* mosquitoes. According to Kim et al. (2002), the biologically active components of the *Foeniculum* fruits have been identified as (+)-fenchone and (*z*)-9-octadecanoic acid.

4.2.8.3 *Antihirustism activity*

It has been observed that *F. vulgare's* ethanolic extract exhibits antihirustism activity. Patients in double-blind research received creams consisting of 1%, 2%, and placebo-containing fennel extract. It is preferable to use a cream with 2% fennel instead of 1% fennel (Javidnia et al., 2003).

4.2.8.4 *Effect on uterine contraction*

There have been reports on how fennel essential oil affects rats' uterine contractions. The intensity of contractions induced by PGE2 and oxytocin was significantly reduced when fennel essential oil at varying doses was administered (10 and 20 µg/ml for PGE2, and 25 and 50 µg/mL for oxytocin, respectively). The frequency of contractions caused by PGE2 was also decreased by fennel essential oil, but not by oxytocin. Using the moving average approach, the estimated LD50 in female rats was 1,326 mg/kg. Moreover, the vital organs of the deceased animals showed no signs of damage (Ostad et al., 2001).

4.2.8.5 *Human liver cytochrome P450 3A4 inhibitory activity*

It has been discovered that 13 substances separated from the fennel methanolic extract have inhibitory effects on human liver cytochrome P450 3A4. With an IC_{50} value of 18.3 lM and a mixed type of inhibition, 5-MoP exhibited the strongest inhibition among these compounds (Figure 4.5) (Subehan et al., 2007).

4.2.8.6 Potential mechanism

Table 4.1 illustrates a few possible mechanisms of the plant's effects.

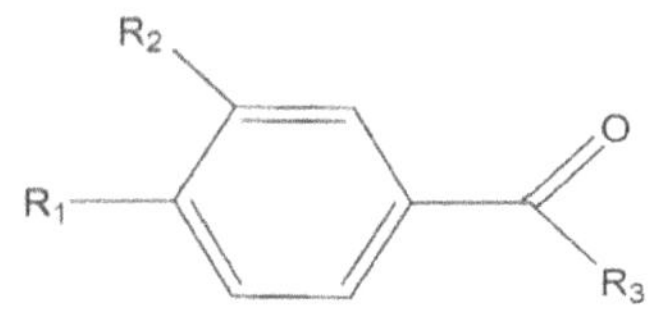

R_1	R_2	R_3	
OCH_3	H	CH_3	Methyl 4-methoxy benzoate
OCH_3	H	H	4-Methoxy benzaldehyde
OCH_3	H	OH	4-Methoxy benzoic acid
OCH_3	OCH_3	OH	3,4-Dimethoxy benzoic acid
OH	OCH_3	OH	4-hydroxy-3-methoxybenzoic acid

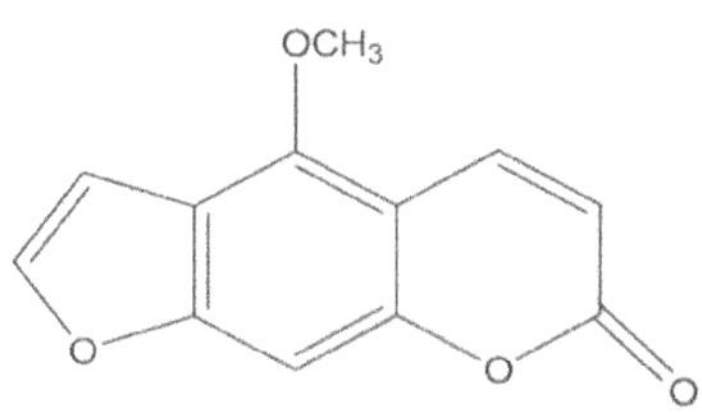

5-methoxypsoralen (5-MOP)

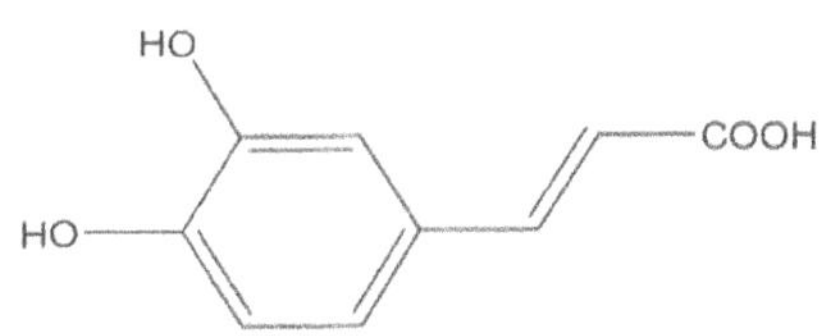

Caffeic acid

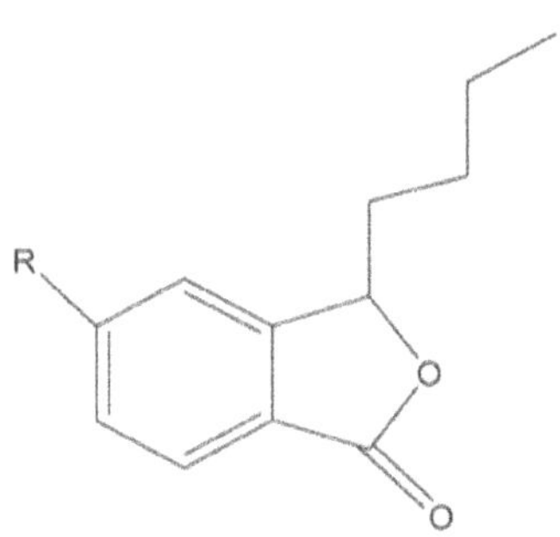

R = OH, 3-butyl-5-hydroxyphthalate

R = OCH_3, 3-butyl-5-methoxyphthalate

Figure 4.5 Molecular structures of bioactive compounds from the methanol extract of *F. vulgare* seeds.

Table 4.1 The potential mechanisms underlying the effectiveness of certain pharmacological properties of *F. vulgare* mill

Pharmacological effects	Possible mechanisms of efficacy
Anti-microbial activity	Existence of antimicrobial compounds in both alcoholic and aqueous extracts, such as coumarin and oleic acid
Antioxidant activity	Antioxidant compounds are found in ethanolic and aqueous extracts, such as phenols and flavonoids
Anti-inflammatory activity	Methanol extract's ability to inhibit lipoxygenase and cyclooxygenase can prevent type 4 allergic reactions and acute and subacute illnesses
Anti-anxiety activity	Effects of anxiety reduction mediated by estrogen and GABA receptors
Gastro-protective activity	Control over the movement of intestinal muscles, management of gastrointestinal spasms and chronic colitis, prevention of gastric ulcers, and thinning of the stomach mucosa
Estrogenic activity	Compounds like the anethole effect on milk secretion can help with infertility, primary dysmenorrhea, menstrual pain, and delivery
Anti-lipid activity	Reduction of plasma triglycerides, LDL, total cholesterol, increased HDL, decreased apolipoprotein B, and increased apolipoprotein A-1, resulting in hypolipidemic effects
Cardiovascular activity	Reduction in SBP and in the amount of water, potassium, and sodium excreted
Anti-diabetic activity	The hypoglycemia effect is characterized by decreased blood glucose, elevated liver and muscle glycogen levels, and increased glutamine peroxide activity
Anti-cancer activity	Active substances like anethole, which have an inhibitory effect on TNF-α activation, antiangiogenic, apoptotic, and antitumor effects, are present
Hepato-protective activity	Decreased levels of proinflammatory cytokines like TNF-α and CTGF in the fibrosis, as well as decreased levels of AST, ALP, ALT, and bilirubin
Memory-protective activity	Effect of inhibition on the enzyme acetylcholinesterase

4.2.9 Safety

There should be more research done on the safety of spice and medicinal plants, as well as the preparations made from them. A primary prerequisite for the application of herbal remedies in medical settings is the absence of potential hazards such as carcinogenicity, teratogenicity, and mutagenicity. Such products should generally have low levels of toxicity and adverse effects. The vast majority of herbal remedies are generally accepted to be safe, with individual hypersensitivity typically regarded as the most frequent but manageable risk. The question of potential harmful effects in humans remains unanswered for specific compounds that are toxic in laboratory animals. Some compounds, most notably estragole, have come under investigation in the case of *F. vulgare*. Estragole, also known as methylchavicol, is one of the primary ingredients in *F. vulgare* essential oil. There have been reports linking estragole to the growth of cancerous tumors in rodents. This served as the foundation for the Scientific Committee on Food (SCF) of the European Union's recommendations to limit the utilization of this material (European Commission, 2001); however, it is still unknown if estragole can cause human cancer. First reported by Drinkwater et al. (1976), estragole's capacity to cause genotoxicity and subsequent carcinogenesis was later confirmed by several in vitro and in vivo studies (Swanson et al., 1979, 1981; Miller and Miller, 1983; Miller et al., 1983; Paini et al., 2010). It was discovered that estragole has carcinogenic effects that are specific to certain tissues, species, and sexes. Recent data suggests that estragole does not directly cause cancer. Rather, its metabolic activation plays a central role in its carcinogenicity; this activation produces unstable molecules and active radicals that adduct with nucleic acids, ultimately damaging DNA (Phillips, 1994). Higher doses of estragole promote its biotransformation, which leads to the production of mutagenic metabolites (Punt et al., 2008). Estragole metabolism (Figure 4.6) is therefore dosage-dependent. The question of whether humans produce estragole's mutagenic metabolites arises from the fact that animals and humans can undergo different biotransformation pathways for the same.

4.2.10 Extraction of antioxidants

Using hydro-distillation with varying distillation times (30 minutes, 1, 2, and 3 hours), essential oils were extracted from both the commercial aerial parts and fruits of *F. vulgare*. These oils were then analyzed using GC and GC-MS. Four different approaches were used to assess their antioxidant properties, and agar diffusion was employed to measure antibacterial activity. Significant variations were found in the essential oils, raising concerns about quality and safety. Methyl chavicol (estragole) (79%–88%)

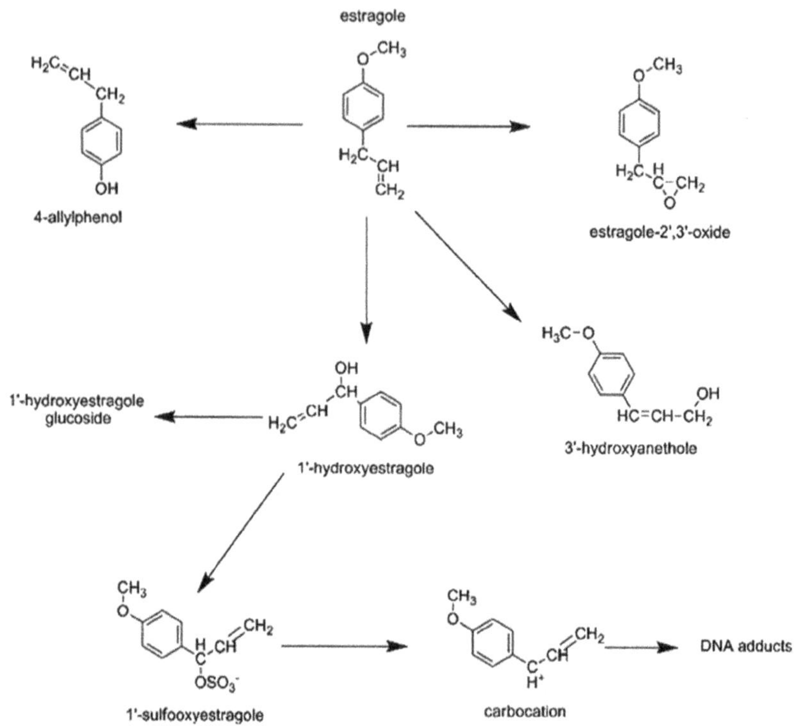

Figure 4.6 Metabolism of estragole. (Adapted from Punt et al. (2008).)

predominated in the fruit oils, while α-pinene (14%–20%), trans-anethole (31%–36%), and limonene (11%–13%) were the primary constituents of the essential oils extracted from the dried aerial parts of *F. vulgare*. The oils from the aerial parts exhibited superior antioxidant activity when tested using the DPPH method. However, fennel essential oils at higher concentrations showed pro-oxidant activity when analyzed using the TBARS method. All oils demonstrated the ability to inhibit 5-lipoxygenase, but none were effective in removing hydroxyl radicals beyond 50%. The antibacterial activity of the essential oils was notably weak. Oils isolated after 2 hours of distillation generally exhibited similar biological activity to those isolated after 3 hours (Miguel et al., 2010).

The antioxidant activity of water along with the ethanol extracts of fennel (*F. vulgare*) seed (FS) was evaluated using many antioxidant assays, involving total free radical scavenging, antioxidant, hydrogen peroxide scavenging, superoxide anion radical scavenging, metal chelating activities, and reducing power. The standard antioxidants' activities, which include BHA, BHT, and α-tocopherol, were compared to those of the other antioxidants. Strong antioxidant activity was demonstrated by the FS seed extracts in both water and ethanol. When evaluated in the linoleic acid system, the water and ethanol extracts, at a dosage of 100 mg, exhibited inhibition of peroxidation at rates of 99.1% and 77.5%, respectively, surpassing the peroxidation inhibition rate of α-tocopherol (36.1%) at the same dosage. The two FS extracts demonstrated free radical scavenging, effective reducing power, hydrogen peroxide scavenging, superoxide anion radical scavenging, and metal chelation activities. This antioxidant characteristic is concentration-dependent and increases with sample volume. Furthermore, the gallic acid equivalents of the total phenolic compounds in the fennel seed water and ethanol extracts were calculated. The current study's findings suggest that fennel (*F. vulgare*) seeds may be a good source of free radicals. Even though the tests conducted here demonstrate the value of FS extracts as in vitro antioxidants, these extracts still need to demonstrate their activity in biological systems, emulsions, health implications, or dry foods (Oktay et al., 2003).

Foeniculum vulgare volatile oil was subjected to GC and GC–MS analysis, which showed the existence of 35 components that accounted for 96.4% of the total. Trans-anethole made up the majority (70.1%). Nine different components made up 68.9% of the total amount, according to the analysis of the acetone extract. The extract's main constituents have been observed to be oleic acid (5.4%), palmitic acid (5.4%), and linoleic acid (54.9%). Utilizing various methods, the antifungal along with antioxidative potentials have been also evaluated. At a 6 lL dose, the volatile oil completely reduced the growth of *A. niger, A. flavus, F. graminearum,* and *F. moniliforme* in the inverted petriplate method. Even at a 4 lL dosage, it was found to be effective against *A. niger*. Furthermore, the volatile oil along with extract both displayed a good to moderate ZOI when tested using the food poisoning technique. The peroxide and thiobarbituric acid values for linseed oil were measured at predetermined intervals to evaluate the antioxidant value. Comparing the volatile oil and extract to BHA and BHT, both demonstrated significant antioxidant activity. Additionally, using the ferric thiocyanate method to track peroxide accumulation in the emulsion at the time of incubation, their inhibitory action in the linoleic acid system has been investigated (Singh et al., 2006).

The chemical composition as well as antioxidative activity of Moroccan *Foeniculum vulgare* Mill. essential oils and solvent extracts were assessed. Gas chromatography-mass spectral (GC-MS) and Gas chromatography (GC) have been utilized to analyze *F. vulgare*. β-pinene (17.8%) and limonene (20.8%) were the principal constituents, with myrcene (15.0%) and fenchone (12.5%) following closely behind. Using a DPPH test system, this aromatic plant antioxidant activity has been assessed in order to propose it as a novel candidate for natural antioxidant sourcing. Spectrophotometry was used to determine the concentrations of phenolic along with the flavonoid compounds in solvent extracts (diethyl ether and ethyl acetate).

Additionally, the extracts' ability to scavenge DPPH increased in the following order: ascorbic acid < diethyl ether extract < essential oil < ethyl acetate extract. Lastly, an association among the extract's total phenolic as well as flavonoid levels and its antioxidant activity potential was found (El Ouariachi et al., 2014).

F. vulgare is a fragrant, medicinal plant having a broad variety of pharmacological effects that are especially significant to the food industry. Anethole (and its polymers, dianethole and photoanethole), (+)-fenchone, estragole, and *P*-anisaldehyde are among its aroma-active compounds that have been identified as biologically active molecules with acaricidal, oestrogenic, and antithrombotic properties. Numerous studies have demonstrated the strong antioxidant potential of the phenolic molecules found in fennel. If the pharmacokinetics, bioavailability, physiological pathways, and significance to human health of these bioactive compounds in fennel are sufficiently understood, they may be formed as unique pharmacological lead molecules.

4.2.11 *Carcum carvi*

Grown as a biennial plant in western Asia, Northern Africa, and Europe, caraway (*Carum carvi*) [black zeera] is a member of the Apiaceae family (Mardani et al., 2016; Keshavarz et al., 2013). Persian cumin is another name for it. It is an umbelliferous, aromatic, biennial plant. The plant grows to a height of 20–30 cm (Miraj and Kiani, 2016). Little white or pink flowers arranged in umbels make up the main flower stem. Caraway fruits are 2 mm long, crescent-shaped achenes. Heartburn, gas, bloating, mild spasms, and loss of appetite in the intestines and stomach are among the digestive issues that caraway can help with (Saghir et al., 2012). Additionally, caraway oil is used to treat constipation, enhance urine control, help people cough up phlegm, and destroy bacteria in the body. Caraway oil is used by women to induce menstruation and ease cramps; it is also used by nursing mothers to boost milk production (Sofi et al., 2009). Mouthwashes and skin rubs containing caraway are used to increase local blood flow. *Carum carvi*, or caraway, seeds have been utilized to treat a variety of conditions, including diarrhea, stomatitis, as a stimulant, hysteria, dyspepsia, colic, dyspeptic headache, flatulent indigestion, and liver function enhancement. They are also used to treat gastrointestinal issues and obesity, as well as having anticonvulsant, reno-protective, anti-hyperglycemic, and antimicrobial properties (Kazemipoor et al., 2014). The herb *Carum carvi* is widely available and has long been utilized in conventional medicine as a medication and as a food additive (Aćimović et al., 2016). According to Subbaraj et al. (2013), it may have biological impacts that include analgesic, anti-obesity, anti-anxiety, anti-inflammatory, and antispasmodic properties, as well as treating hypothyroidism.

Caraway, or Carumcarvi L. [Apiaceae family], is a common houseplant grown all over the world, including in Iran. Caraway fruits have several traditional uses in ethnomedicine and are used as flavoring in food and drink. Caraway's spasmolytic, anti-inflammatory, antioxidant, carminative, antimicrobial, and immunomodulatory qualities suggest that it may help treat inflammatory bowel disease (IBD) (Charles, 2012).

4.2.12 Nutritional uses

Caraway seeds are enriched with many important nutrients, such as vitamins A, B-complex, C, and E (Table 4.2). These vitamins are rich in health-promoting qualities that are essential for promoting overall body wellness. The significant amounts of minerals that caraway seeds contain, including calcium, magnesium, copper, iron, zinc, potassium, selenium, and manganese, add to their benefits. Each of these minerals serves a distinct health purpose to support and maintain our body's overall wellness and fitness. The high dietary fiber content of caraway seeds is another factor contributing to their health benefits.

Table 4.2 Nutrient composition of caraway seeds

Nutrients	Units	Values per 100 g
Water	G	9.87
Protein	G	19.77
Energy	Kcal	333
Total lipid (fat)	G	14.59
Fiber, total dietary	G	38.0
Carbohydrate, by difference	G	49.90
Calcium, Ca	mg	689
Sugars, total	G	0.64
Vitamin C, total ascorbic acid	mg	21.0
Vitamin B-12	mcg	0.00
Vitamin B-6	mg	0.360
Vitamin A, RAE	Mcg-RAE	18
Vitamin D	IU	0
Fatty acids, total saturated	G	0.62
Vitamin E (alpha-tocopherol)	mg	2.5
Vitamin A, IU	IU	363
Fatty acids, total monounsaturated	G	7.125
Fatty acids, polyunsaturated"	G	3.272

Source: USDA (2011).

4.2.12.1 Chemical compounds

Major phytochemicals found in *C. carvi* seeds include p-cymene, γ-terpinene, α-pinene, limonene, carvacrol, carvone, and carvenone (Figure 4.7) (Esmaeili and Asgari, 2015). According to Baananou et al. (2013), the essential oil compositions of various plant parts of wild caraway contain a higher proportion of limonene (61%–83%), while those from the inflorescences contain germacrene D (23%–41%) and limonene (39%–62%). Germacrene D and germacrene A dominated the low-volatile leaves and stems (Miraj and Kiani, 2016). High levels of carvone and limonene have been observed in the essential oils extracted under SFE conditions (Baananou et al., 2013). Young fruit oils contained a high percentage of limonene (61%–83%), while inflorescence oils contained limonene (39%–62%) and germacrene D (23%–41%). Fruits, inflorescences, and leaves all had varying amounts of carvone. Germacrene D and A predominated in the low-volatility leaves and stems (Kazemipoor et al., 2014).

4.2.12.2 Appetite-suppressing activity

The appetite-suppressing properties of CAE (Caraway Aqueous Extract) were investigated in 70 aerobically trained, obese, and overweight women in an in vivo study. For 90 days, subjects were randomly assigned to experimental and placebo groups, and they were given either 30 mL/day of CAE or a placebo without altering their diet or level of physical activity. The experimental group's carbohydrate intake and appetite levels were significantly lower than those of the placebo group, according to the results. According to Kazemipoor et al. (2014), these early results imply that a

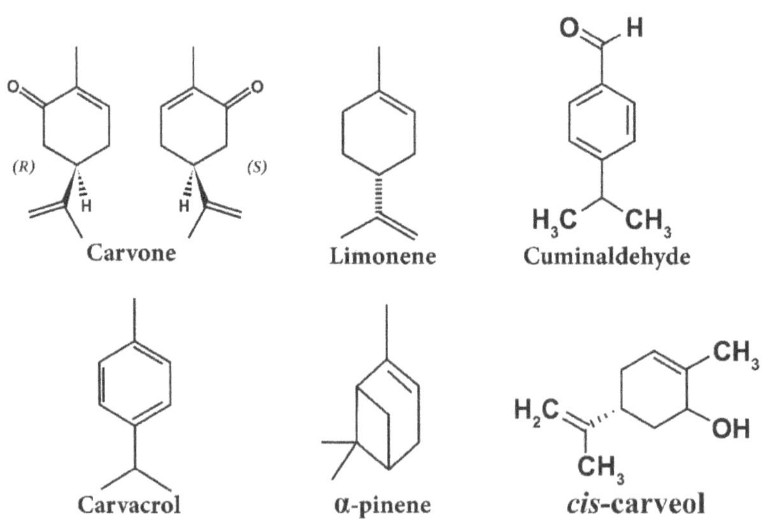

Figure 4.7 Structures of some phytoconstituents isolated from *Carcum carvi*.

dietary CAE may be useful in helping physically active adult females control their weight by decreasing their hunger levels and body size.

4.2.12.3 *Oral mucositis*

Animal research has been performed to examine the impact of caraway hydroalcoholic extract on 5-fluorouracil-induced oral mucositis (OM). The study revealed that caraway exhibited greater efficacy against *Streptococcus intermedius* and *Staphylococcus epidermidis*. γ-Terpinene has been identified as the main component in the essential oil. This extract can be topically applied to reduce the intensity of OM because of its antibacterial activity along with its terpinene content (Mardani et al., 2016).

4.2.12.4 *Irritable bowel syndrome*

When caraway oil poultices were evaluated for their effectiveness in treating IBS, it was discovered that they were effective and side-effect-free. However, because of their heat therapy properties, they may still be beneficial. They are therefore suitable and helpful in the self-management of IBS (Lauche et al., 2015).

4.2.12.5 *Hypothyroidism*

Case research described a patient with papillary thyroid carcinoma who was receiving levothyroxine treatment and developed increased TSH level as a result of consuming *Carum carvi*. According to Naghibi et al. (2015), the study's findings indicated that *Carum carvi* consumption may raise TSH levels in hypothyroid patients receiving levothyroxine. A case study described a patient with papillary thyroid carcinoma who was receiving levothyroxine treatment as well as developed a raised level of TSH as a result of consuming *Carum carvi*. According to Naghibi et al. (2015), the study's findings indicated that *Carum carvi* consumption may raise TSH levels in hypothyroid patients receiving levothyroxine.

4.2.12.6 *Blood pressure lowering, vasodilator and cardiac-modulatory*

Caraway was found to have independent, endothelium-dependent, and BP-lowering cardiomodulatory and vasodilator properties in animal research. The findings suggested that the seed extract's cardiovascular effects may be mediated by a combination of phosphodiesterase inhibitory, nitric oxide modulating, and Ca++ antagonist mechanisms. Its exact mode of action is still unknown, though (Miraj and Kiani, 2016).

4.2.12.7 *Antitubercular activity*

Carum carvi's effectiveness in modifying the pharmacokinetics of isoniazid, pyrazinamide, and rifampicin in a fixed dosage combination has been assessed in a human study. The results indicated that adding *Carum carvi* causes an increase in plasma levels of these three drugs, which in turn allows it to act as a bioenhancer and positively alter the antitubercular treatment kinetics (Choudhary et al., 2014).

4.2.12.8 Anti-cancer

The impact of dietary *caraway* oils on cancer progression was investigated in an animal study, focusing on the β-catenin expression in the colon at the time of DMH-induced colonic carcinogenesis. According to Allameh et al. (2013), there has been evidence that the Wnt/β-catenin signaling pathway is triggered at the time of the promotion of colon cancer and that long-term *caraway* oil feeding modifies the colonic β-catenin expression, which in turn suppresses DMH-induced premalignant lesions in the rat colon.

Two TNBC cell lines with mutant p53 were used to test TQ's growth inhibitory effect. Cell metabolism experiments demonstrated that TQ increased the cytotoxicity induced by docetaxel and cisplatin while inhibiting the growth of TNBC cells without influencing normal cell growth. Accordingly, it is hypothesized that TQ may be helpful in the treatment of TNBC even in the absence of functional p53.

4.2.12.9 Anti-colitis activity

The effects of caraway hydroalcoholic extract (CHE) and its essential oil (CEO) in an immunological model of colitis in rats induced by trinitrobenzene sulfonic acid (TNBS) were investigated. It was found that caraway fractions are both effective and possess anti-colitis activity irrespective of the dose and route of administration (Keshavarz et al., 2013).

4.2.12.10 Cardiovascular disease

The caraway plant's effects on lipid profiles, levels of blood glucose, and CRP in diabetic rats were examined in an animal study. *Carum carvi* might have a reducing effect on blood pressure and cholesterol levels in diabetics while having no effect on the level of CRP (Sadjadi et al., 2014),

4.2.12.11 Anti-obesity activity

Researchers looked into how CE affected physically active, obese, and overweight women's weight. Subjects did not alter their diet or level of physical activity; they were given a placebo or 30 mL of CE per day. The subjects' blood pressure, urine-specific gravity, and lipid profile did not change. The findings imply that a dietary CE without any food restrictions, when paired with physical activity, can help women who want to reduce their body fat percentage, weight, BMI, and size manage their obesity. These study outcomes point to a potential use of *caraway* extract as a phytotherapeutic strategy for obesity treatment (Kazemipoor et al., 2014).

4.2.12.12 *Antimicrobial activity*

Pullulan films infused with caraway essential oil were tested for their capability to inhibit the growth of *Aspergillus niger, Saccharomyces cerevisiae, Staphylococcus aureus,* and *Salmonella enteritidis*. It was revealed that, of the tested species, *S. enteritidis* exhibited the highest level of resistance. According to the research results, it might be possible to extend the microbiological stability of minimally processed foods by applying a pullulan film that has CEO integrated (Gniewosz et al., 2013).

4.2.12.13 *Hypolipidemic effect*

Rats with diet-induced hyperlipidemia were evaluated for the impact of *Carum carvi* seed aqueous extract; the results showed that this plant's aqueous seed extract decreased lipid levels in rats with diet-induced hyperlipidemia (Saghir et al., 2012).

4.2.12.14 *Antihyperglycemic and hypolipidemic activity*

An assessment was conducted on the weight, lipid profile, and blood glucose level of diabetic rats following oral administration of caraway. The findings demonstrated that giving caraway orally to treated rats significantly decreased their blood glucose levels and lessened their body weight loss. In diabetic rats, caraway demonstrates both hypolipidemic and antihyperglycemic effects. However, to use it in real life, more in vivo research is necessary (Gniewosz et al., 2013).

4.2.12.15 *Anti-oxidative stress activity*

Caraway extracts have been studied in animals to see if they can stop sepsis caused by oxidative tissue injuries. The induction of sepsis resulted in a noteworthy elevation of kidney lactate production, while there was no discernible effect on heart lactate production. Rats treated with essential oils showed an easy reversal of kidney LP and plasma urea/creatinine ratio levels; rats treated with hydroalcoholic extract did not show the same response. Consequently, it was demonstrated that in advanced stages of sepsis, caraway oils likely play a protective role in kidney tissue against oxidative injury (Miraj and Kiani, 2016).

4.2.12.16 *Potential side effects*

Oral consumption of CAE has been determined to be safe when taken at a dosage of 30 mL/day for a duration of 12 weeks (Miraj and Kiani, 2016).

4.2.13 *Extraction of antioxidants*

The antimicrobial, antioxidant, anti-acetylcholinesterase, and antidiabetic properties of the EOs extracted from the seeds of *C. sativum* and *C. carvi* were assessed chemically, both separately and in combination.

The majority of the EOs were found to be monoterpenes, with the main constituents of *C. carvi* EO being β-pinene (18.77%), γ-terpinene (31.03%), carvone (12.20%), and p-cymene (17.16%), and *C. sativum* EO consisting of linalool (76.41%), γ-terpinene (5.35%), and α-pinene (4.44%). Statistical analysis demonstrated that *C. carvi* EO had higher and significantly different ($p < 0.05$) antioxidant activity than *C. sativum* EO, but lower than the mixture, when compared to standards. In addition, the mixture demonstrated two times higher equipotent chelating power (IC50 = 31.33 ± 0.47 mg/mL) and ferric ion reducing antioxidant power (FRAP) (IC50 = 11.33 ± 1.53 mg/mL) than the corresponding references. Additionally, it demonstrated strong activity against assays measuring DPPH (IC50 = 19.00 ± 1.00 mg/mL), β-carotene (IC50 = 11.16 ± 0.84 mg/mL), and superoxide anion (IC50 = 10.33 ± 0.58 mg/mL). According to antimicrobial data, both single and mixture essential oils (EOs) exhibited activity against a range of pathogenic microorganisms. The mixture proved to be more effective in killing bacterial strains than either EO alone. Furthermore, the first study of the anti-acetylcholinesterase and α-glucosidase inhibitory effect has been conducted. The results show that *C. sativum* (IC50 = 0.68 ± 0.03 mg/mL) and *C. carvi* (IC50 = 0.82 ± 0.05 mg/mL) have a high inhibitory effect on AChE, as does the mixture (IC50 = 0.63 ± 0.02 mg/mL) when compared to the reference drug, which is not statistically different ($p > 0.05$). Comparing the mixture to the standard medication, acarbose, revealed a high and equipotent antidiabetic activity (IC50 = 0.75 ± 0.15 mg/mL), which is approximately nine times higher than each EO alone. In addition, the utilization of a blend of *C. carvi* and *C. sativum* essential oils, along with pharmacokinetic analysis, provides valuable insights for the development of novel drugs that possess desirable drug-likeness and safety characteristics. Overall, the study's results suggest that the combination of these EOs could be recommended for further use in food, medicine, and pharmaceuticals. Additionally, it can serve as a therapeutic intervention to proactively mitigate various ailments (Hajlaoui et al., 2021).

The phytochemical composition and biological properties of essential oil (EO) extracted from irradiated caraway seeds (*Carum carvi* L.) were reported by Aly et al. (2023). After gamma irradiation at dose levels of 2.5, 5, and 10 kGy was applied to *Carum carvi* L. seeds, EOs were extracted from each sample. Evaluations were done on the composition of chemicals, phenol content, and flavonoid content. Investigations were also conducted into antitumor and antimicrobial activities against two cell lines—liver cancer (HepG-2) and colonic adenocarcinoma (Caco-2). According to the results, radiation increased the percentage of oil, with a dose of 10 kGy giving the greatest percentage (3.50%) compared to the control. Additionally, the essential oil extracts contained 26 compounds, according to the GC-MS analysis. Carvone was the primary ingredient in caraway seed EO, followed by limonene. In comparison to the control, the results depicted that using gamma rays raised the content of flavonoids and phenols. The maximum increase was noted at dosage levels of 10 kGy (13.70 and 7.38 mg/g oil, respectively), followed by 5 kGy (11.20 and 5.86 mg/g oil, respectively). The caraway essential oils' antioxidant properties were enhanced through an increase in irradiation dosage level (2.5–10 kGy), as determined by metal chelating activity and DPPH radical analysis. Essential oils of caraway possess antimicrobial properties that can combat both Gram-positive and Gram-negative bacteria, as well as fungi. The 10 kGy dosage was more successful in inhibiting the growth of fungi and bacteria, and the antimicrobial activity increased as the irradiation dosage was raised. Furthermore, as demonstrated by decreased cell viability and density, the caraway EOs exhibit anticancer activity against the two cell lines under investigation: the liver cancer cell line (HepG-2) and colorectal adenocarcinoma (Caco-2). In terms of oil yield, flavonoids, phenols, antibacterial and antifungal properties, and antioxidant activity, the 10 kGy dose demonstrated superior efficacy. Additionally, as cell viability and density were reduced in the two cell lines under investigation—colorectal adenocarcinoma (Caco-2) and liver cancer (HepG-2)—the caraway essential oils showed anticancer activity. Thus, caraway may be regarded as a significant herb with a variety of medicinal applications.

The seeds of *C. carvi* are a rich source of essential oils that contain a wide range of phytochemicals. They have a broad range of pharmacological effects when used in traditional healing systems around the globe. In order to fully explore the pharmacological action of *C. carvi* seeds, numerous studies have been conducted over the past 20 years.

4.2.14 *Pimpinella anisum*

Among the earliest known medicinal plants is *P. anisum* L., a member of the Umbelliferae family. The plant thrives in the Eastern Mediterranean region, the Middle East, West Asia, Egypt, Mexico, and Spain. It is a perennial herbaceous plant that reaches a height of 30–50 cm and produces white flowers along with small seeds that range in color from green to yellow (Salehi Surmaghi, 2010). *P. anisum* is primarily cultivated for its fruits, known as aniseeds, which are harvested during the months of August and September. Aniseeds are employed as a seasoning, digestive aid, carminative, and for alleviating gastrointestinal spasms. Their composition consists of 1.5%–5% EO. Anise seed consumption increases milk production in nursing mothers and helps ease gastrointestinal issues in their infants (Zargari, 1996). Anise is used as an aromatic and flavoring ingredient in fish products, ice cream, candies, and gums in the food industry (Ozcan and Chalchat, 2006). There isn't a thorough review study on the components and efficacy of *Pimpinella anisum*, despite the numerous studies that have been done on it.

4.2.15 *Pimpinella anisum in traditional medicine*

In conventional medicine, anise seeds have been utilized as a carminative, aromatic, diuretic, and migraine analgesic (Amin, 2005). Warm and dry in nature, aniseed promotes the production of milk, menstruation, urine, sweat, and a healthy complexion. It also works well for tooth polishing. Anise is mentioned in some traditional texts as a remedy for depression, nightmares, and seizures, as well as epilepsy (Mirheydar, 2001).

4.2.16 *Chemical constituents*

Aniseed comprises a volatile oil, predominantly trans-anethole, constituting 1.5%–6.0% of its total weight. The composition of this substance includes lipids that are abundant in fatty acids, specifically oleic and palmitic acids, constituting approximately 8%–11% of its total weight. Moreover, aniseed comprises approximately 4% carbohydrates and 18% protein, as reported by Besharati-Seidani et al. (2005). The main constituents of anise seed essential oil have been shown in other investigations to include *eugenol, trans-anethole, estragole, methylchavicol, coumarins, scopoletin, anisaldehyde, umbelliferone, estrols, polyenes, polyacetylenes, and terpene hydrocarbons* (Gulcin et al., 2003). Estragole (2.4%) and trans-anethole (93.9%), as determined by GC and GC-MS analysis of the EO of *Pimpinella anisum* L. fruits, have been identified. Ozcan and Chalchat (2006) reported the presence of additional compounds with concentrations greater than 0.06%, including (*E*)-methyleugenol, α-himachalene, α-cuparene, β-bisabolene, cis-anethole, and *p*-anisaldehyde. Other compounds like p-anisaldehyde, trans-pseudoisoeugenyl 2-methylbutyrate, and methylchavicol have also been identified in the EO of *P. anisum* L. fruits obtained from Europe's geographical areas, in addition to the major components (trans-anethole, 76.9%–93.7%) and γ-himachalene (0.4%–8.2%) (Orav et al., 2008). An analysis of the constituents of the whole plants and the seeds of *Pimpinella anisum* from Alberta revealed that trans-anethole accounted for 75.2% of the seed oil and 57.4% of the whole plant. Embong et al. (1997) identified the remaining components of the plant oil as carvone, cis-anethole, dihydrocarvyl acetate, β-caryophyllene, estragole, and limonene, which were present in amounts ranging from 1% to 5%.

GC-MS has been utilized to identify the chemical components of the aniseed extract that was produced through supercritical extraction with CO_2. According to Rodrigues et al. (2003), the main constituents were anethole (approximately 90%), *p*-anisaldehyde (<1%), γ-himachalene (2%–4%), cis–pseudoisoeugenyl 2-methylbutyrate (~3%), methylchavicol (0.9%–1.5%), and trans–pseudoisoeugenyl 2-methylbutyrate (1.3%). In 1978, neophytadiene, a novel terpene hydrocarbon, was discovered in aniseed (Burkhardt et al., 1986). Aniseed was also used to isolate 4-(β-D-glucopyranosyloxy) benzoic acid, a phenolic glycoside belonging to the Umbelliferae family (Dirks and Herrmann, 1984). Four aromatic compound glucosides, an alkyl glucoside, and a glucide were identified as novel compounds in a study by Fujimatu et al. (2003) from the polar part of the anise fruit methanolic extract. (*E*)-3-Hydroxy-anethole β-D-glucopyranoside, 3-hydroxyestragole β-D-glucopyranoside, (*E*)-10-(2-hydroxy-5-methoxyphenyl) propane β-D-glucopyranoside, hexane-1,5-diol 1-*O*-β-D-glucopyranoside, methyl syringate 4-*O*-β-D-glucopyranoside, and 1-deoxy-L-erythritol 3-*O*-β-D-glucopyranoside were the clarified structures of the new compounds. Separation and organization using chromatography on cellulose columns to isolate the flavonoid constituents from caraway, coriander, anise, and fennel, researchers were able to isolate the crystalline compounds rutin, quercetin 3-glucuronide, isoorientin, luteolin 7-glucoside, and isovitexin from the noncrystalline compounds apigenin 7-glucoside along with the luteolin glycoside" (Kunzemann and Herrmann, 1977).

A silver ion HPLC method was employed in a different study to ascertain the aniseed oil's fatty acid composition. The findings demonstrated that in aniseed oil, a single gradient run on a single cation exchange column laboratory converted the positionally isomeric 18:1 fatty acid oleic acid (cis 9–18:1), cis-vaccenic acid (cis 11–18:1), and petroselinic acid (cis 6–18:1) to the silver ion form (Denev et al., 2011). Furthermore, by employing a combination of anion-exchange, gel filtration, and hydrophobic interaction column chromatographies, researchers were able to isolate three complexes consisting of lignin, carbohydrate, and protein from a hot water extract of *Pimpinella anisum* seeds (Lee et al., 2011).

4.2.17 Pharmacological properties

The disc diffusion approach was utilized to test the antibacterial and antifungal properties of the aqueous, 50% (v/v) methanol, petroleum ether, and acetone extracts of *P. anisum* L. fruits against four pathogenic bacteria, namely *Escherichia coli, Streptococcus pyogenes, Klebsiella pneumoniae,* and *Staphylococcus aureus*. The findings depicted that while petroleum and acetone ether extracts were unable to stop the growth of the pathogenic test bacteria, only aqueous and methanol extracts demonstrated moderate antibacterial activity against all of the test bacteria, with the aqueous extract outperforming the methanolic extract (Akhtar et al., 2008).

Gulcin et al. (2003) investigated the antimicrobial properties of water and ethanolic aniseed extracts against ten different bacterial species, along with *Candida albicans*, using the disc diffusion approach. In this investigation, the ethanolic extract was ineffective against *C. albicans* but demonstrated strong inhibitory activity against all other tested bacteria. The water extract had antimicrobial activity against *C. albicans* but not against *E. coli, Pseudomonas aeruginosa*, or Gram-negative bacteria. Ateş and Turgay (2003) found that *Pimpinella anisum* seed alcoholic extracts inhibited *Micrococcus luteus* and *Mycobacterium smegmatis*.

In different research, the combined antibacterial activity of methanol extract and essential oils of *Thymus vulgaris* and *Pimpinella anisum* has been assessed against nine pathogenic bacteria. The essential oil and methanol extract of these plants demonstrated antibacterial activity against the majority of tested pathogens, with *Bacillus cereus, Staphylococcus aureus,* and *Proteus vulgaris* showing the greatest response. However, against the majority of tested bacteria, particularly *Pseudomonas aeruginosa*, the essential oil and methanol extracts of these plants demonstrated an additive effect (Al-Bayati, 2008). Aqueous decoctions of black pepper, bay leaf, aniseed, and coriander were tested for antibacterial activity against 176 bacterial isolates from 12 genera using the mean of disc diffusion. The anise decoction exhibited 18.1% antibacterial activity, peaking against *Micrococcus roseus*. On the other hand, the black pepper aqueous decoction demonstrated an antibacterial activity of 75%, making it the most effective in killing bacteria (Chaudhry and Tariq, 2006). Anethole was the most active ingredient in anise essential oil, which demonstrated strong inhibitory activity against fungi in addition to its antibacterial properties (Shukla and Tripathi, 1987). The antifungal properties of anise fruit fluid extract and essential oil have been examined in research by Kosalec et al. (2005) using the broth dilution approach and the diffusion method with cylinders on seven species of yeasts and four species of dermatophytes. The results showed that *C. albicans, C. parapsilosis, C. pseudotropicalis, C. tropicalis,* and *C. krusei* were all susceptible to the antimycotic effects of aniseed fluid extract, with *Candida albicans* exhibiting the largest inhibition zone. Additionally, it demonstrated an inhibitory effect on the dermatophyte species *T. mentagrophytes, Trichophyton rubrum, M. gypseum,* and *Microsporum canis*, with *T. mentagrophytes* exhibiting the largest inhibitory zone. Strong antifungal activity of anise EO was observed against yeasts as well as dermatophytes; *C. parapsilosis* showed the largest inhibition zone (30 mm), followed by *C. albicans, C. glabrata,* and *Geotrichum* spp. According to Kosalec et al. (2005), anise essential oil demonstrated more potent antifungal properties in this investigation compared to its extract against dermatophytes and yeasts. *Alternaria alternata, Aspergillus niger,* and *Aspergillus parasiticus* have all been shown to be susceptible to the antifungal properties of anise essential oil, with *A. parasiticus* being the most affected (Ozcan and Chalchat, 2006). Aniseed methanolic extracts were tested for their antifungal properties against four dermatophyte species, along with one saprophytic fungus. The results indicated that the extract only inhibited the dermatophyte species, such as *Trichophyton mentagrophyte, Microsporum canis,* and *Candida albicans*, but had no impact on the saprophyte *Aspergillus niger* (Yazdani et al., 2009).

4.2.17.1 *Muscle relaxant effect*

Boskabady and Ramazani-Assari (2001) investigated the relaxant effect of *P. anisum* on isolated guinea pig tracheal chains and its potential mechanism. This study examined the bronchodilatory impacts of the essential oil (EO) of anise, ethanol, and aqueous extracts on precontracted isolated guinea pig tracheal chains under two distinct conditions: non-incubated tissues (group 1) and incubated tissues with 1 mM propranolol and 1 mM chlorpheniramine (group 2). Theophylline (1 mM), essential oil, and aqueous and ethanol extracts all demonstrated noteworthy relaxant effects in comparison to the controls, according to the findings. Theophylline's relaxant effects were not substantially different from those of aqueous or ethanol extracts, but the essential oil had a much smaller effect than theophylline. Additionally, there were no discernible variations among the relaxant effects from the experiments conducted in groups 1 and 2. The findings also demonstrated that this plant's relaxing properties are a result of its inhibitory actions on muscarinic receptors.

In a different study, rat anococcygeus smooth muscle was used to test the antispasmodic and relaxant properties of *P. anisum* aerial parts' three hydroalcoholic extracts (ethanol:water; 40:60, 60:40, and 80:20). Acetylcholine-induced contraction was attenuated by all three hydroalcoholic extracts. The examination results showed that two of the other hydroalcoholic extracts were unable to induce relaxation, while the extract containing 60% ethanol (HA60) demonstrated concentration-dependent relaxation of acetylcholine-precontracted tissues. Examination of the potential processes that may underlie the relaxant effect revealed that this effect primarily depends on the activation of the NO-cGMP pathway (Tirapelli et al., 2007)".

4.2.17.2 *Anticonvulsant effect*

Male mice were used to test the anticonvulsant properties of an essential oil extracted from the fruits of *P. anisum* against seizures induced by maximal electroshock (MES) or pentylenetetrazole (PTZ). According to this study, *P. anisum* can prevent tonic convulsions caused by an intraperitoneal injection of PTZ and raise the threshold for clonic seizures induced by an intravenous infusion of the drug. Additionally, *P. anisum* exhibits anticonvulsant properties against MES-induced tonic seizures (Pourgholami et al., 1999). Heidari and Ayeli also examined the impact of anise's methyl alcoholic extract on mice experiencing seizures induced by picrotoxin. According to the results, mice pre-treated with varying doses of anise extract beforehand exhibited a greater delayed in the onset of seizures, with 200 mg/kg being the most efficient dose. Furthermore, this dosage was found to postpone death in mice more effectively than phenobarbital (40 mg/kg) (Heidari and Ayeli, 2005). *Rosmarinus officinalis, Matricaria chamomilla*, *P. anisum*, and *Artemisia vulgaris* extracts have been observed "to delay the onset of picrotoxin-induced seizures and reduce the mortality rates (Ghani et al., 1987). These findings were obtained from another study that examined the effects of aqueous leaf extracts and stems from some Arab medicinal plants, including *P. anisum*, on picrotoxin-induced seizures in mice".

Janahmadi et al. (2008) to examine the mechanisms of a cellular responsible for the anise fruit's antiseizure effects. This study examined the impact of anise fruit essential oil on the bioelectrical activity of snail neurons under two conditions: normal control and induced epileptic activity using PTZ. The results indicate that the hyperexcitability caused by anise oil is likely due to either increased activity of Ca^{2+} channels or decreased activity of voltage and/or Ca^{2+} dependent K^+ channels, which are responsible for post-hyperpolarization potential.

4.2.17.3 *Effect on gastrointestinal system*

Effect on Gastric Ulcer: To investigate the impact of an aqueous anise suspension on gastric ulcers in rats, a combination of indomethacin and several harmful chemicals was used to induce acute gastric ulceration. The results showed that anise significantly reduced the damage caused to the stomach mucosa by necrotizing agents and indomethacin. Histological evidence of the antiulcer effect was obtained (Al Mofleh et al., 2007).

Palliation of Nausea: A group of 25 patients experiencing nausea symptoms in a hospice and palliative care program were treated with an aromatherapy combination consisting of *P. anisum*, *Foeniculum vulgare*, *Anthemis nobilis* var. dulce, and *Mentha piperita*. The majority of the patients who underwent aromatherapy treatments reported improved well-being. However, all of the patients in this study were also addressing their symptoms with a range of additional therapies (Gilligan, 2005).

Effect on Constipation: The criteria for chronic constipation, as defined by the American Association of Gastroenterology, were met by 20 patients who participated in a randomized controlled trial (RCT) to assess the laxative effectiveness of a phytotherapy compound consisting of *F. vulgare* Miller, *P. anisum* L., *Cassia angustifolia*, and *Sambucus nigra* L. Colonic transit time (CTT), determined radiologically, was the primary outcome. Secondary endpoints included the number of evacuations per day, perceived bowel function, side effects, and quality of life. When comparing the phytotherapeutic compound's laxative effects to a placebo, the research outcomes were noteworthy. Both a decrease in the time it takes for food to pass through the colon and an increase in the frequency of daily bowel movements were observed as measures of this effect. This compound can serve as a safe alternative for treating constipation, although the quality of life did not show notable differences between study periods, and adverse effects remained minimal throughout the study (Picon et al., 2010).

4.2.17.4 Effect on morphine dependence

Pimpinella anisum essential oil was tested on mice to evaluate its effect on morphine-induced conditioned place preference (CPP) expression and acquisition. The results showed that mice injected with *P. anisum* essential oil exhibited conditioned place aversion, indicating that the oil has aversive effects. Additionally, subcutaneous morphine injections (2–5 mg/kg) produced a dose-dependent place preference. The oil was also found to have GABAergic properties (Sahraei et al., 2002).

4.2.17.5 Analgesic and anti-inflammatory effect

P. anisum and *Tribulus terrestris* extracts showed strong analgesic activity against benzoquinone-induced writhing and in thermal tests, as per several Iranian medicinal plants screening (Twaij et al., 1988). *P. anisum* EO had analgesic properties comparable to aspirin and morphine in a Tas study (Tas, 2009). The anti-inflammatory, along with the analgesic properties of fixed oil of anise, were also studied in mice. According to Tas et al. (2006), the results depicted that the anise fixed oil had an analgesic effect equivalent to 100 mg/kg aspirin and 10 mg/kg morphine at the 30-minute mark, as well as an anti-inflammatory effect comparable to indomethacin.

4.2.17.6 Effect on menopausal hot flashes

Seventy-two postmenopausal women were tested for *anise* extract's effect on hot flashes. This study found that postmenopausal women who take three 100-mg *anise* extract capsules for 4 weeks have significantly fewer hot flashes (Nahidi et al., 2008).

4.2.17.7 Effect on dymenorrhea

Khoda Karami et al. (2008) treated 180 17–28-year-old female students with primary dysmenorrhea with herbal or mefenamic acid capsules containing dried *celery*, *saffron*, and *anise*. Both the herbal and mefenamic acid groups had significantly lower pain intensity than the placebo group. The herbal capsules relieved pain better than mefenamic acid and could replace mefenamic acid for primary dysmenorrhea.

4.2.17.8 Antioxidant activity

Research by Gulcin et al. (2003) compared the antioxidant activities of synthetic antioxidants, including BHT, BHA, and α-tocopherol. Antioxidant tests were performed on *aniseed* water and ethanolic extracts. Both water and ethanolic *aniseed* extracts showed better antioxidant activity, including DPPH radical and superoxide anion scavenging, reducing power, H_2O_2 scavenging, and metal chelation, compared to BHT, α-tocopherol, and BHA.

Investigation of the in vitro and in vivo antioxidant potential of *aniseed* showed that the ethanolic extract of *aniseed* displayed scavenging activity against nitric oxide, superoxide, and 1,1-diphenyl-2-picryl hydrazyl (DPPH) radicals, along with reducing power, in a concentration-dependent manner (Rajeshwari et al., 2011). The antioxidant properties of *anise* seed oleoresins and EOs were also examined. Antioxidant activities were measured by linoleic acid peroxidation inhibition, Fe^{3+} reducing power, lipid peroxidation assays, and DPPH radical scavenging. *Anise* oil and its methanol oleoresin had the highest antioxidant activity, surpassing BHT and BHA. However, some oleoresins exhibited lower antioxidant activity (Singh et al., 2008).

Iranian Umbelliferae fruits, including *P. anisum*, have been tested for antioxidant activity using the DPPH radical scavenging test. All extracts showed antioxidant activity, but the *P. anisum* extract exhibited the highest activity (IC_{50}=109.80) and flavonoid content in the ethyl acetate fraction. Both flavonoid content in the fraction and antioxidant potency were positively correlated. Another study examined the antioxidant activity of water and alcohol extracts of *anise*, dill, and chamomile. The extracts showed antioxidant activity in the linoleic acid and liposome model systems, with water extracts displaying higher antioxidant activity compared to alcohol extracts. Antioxidant activity was lower in *anise* seeds than in dill and chamomile (Mohammad Al-Ismail and Aburjai et al., 2004).

Anise tea was one of 15 herbal teas tested in vitro for antioxidants. *Anise* tea had weak antioxidant activity in the TEAC assay and moderate effects in the hypochlorite and peroxynitrite quenching assays compared to other herbal teas (Speisky et al., 2006).

4.2.17.9 Insecticidal effect

Using a fumigation bioassay, the insecticidal properties of plant EO from 40 species of plants, including *Pimpinella anisum* (PAS), were evaluated against *Lycoriella ingenua* larvae. Several essential oils, such as those derived from garlic and anise, demonstrated effective insecticidal action against larvae. Allyl isothiocyanate has been observed to be the most toxic compound in effective essential oils against larvae of *Leishmania ingenua*, with trans-anethole, *p*-anisaldehyde, and diallyl disulfide following closely behind (Park et al., 2006).

Prajapati et al. (2005) found that *Pimpinella anisum* and *Juniperus macropoda* essential oils were efficient larvicidal and ovicidal against three species of mosquitoes. Anise EO also repels *Culex pipiens* mosquitoes (Erler et al., 2006). After being exposed to anise and cumin essential oils, the *Ephestia kuehniella* (Mediterranean flour moth) and the *Tribolium confusum* (confused flour beetle) lost all their eggs (Tunc et al., 2000).

p-Anisaldehyde and commercial anise seed oil components were tested for their ability to kill house dust mites, *D. pteronyssinus*, and

Dermatophagoides farina. The most toxic compounds to these dermatophagoides have been benzyl benzoate and *p*-anisaldehyde. Thus, *p*-anisaldehyde might be a good lead compound for selective house dust mite control (Lee, 2004).

4.2.17.10 Effect on diabetes patients

Aniseeds, along with the coriander seeds, were then compared for their anti-diabetic, hypolipidemic, and antioxidant properties in patients with type 2 diabetes. For 60 days, two groups of patients with type 2 diabetes received the seed powders (5 g/day). The findings depicted an 11% rise in fasting blood glucose in the control group, a 36% reduction in type 2 diabetics treated with *aniseed*, and a 13% decline in those treated with *coriander*. Furthermore, a noteworthy decrease in serum cholesterol, along with triglycerides, was noted in patients administered with *aniseed* and *coriander* seeds. Lipid peroxidation in plasma and erythrocytes, as well as protein oxidation in serum, were reduced in both treated groups when compared to baseline levels. Increased vitamin A and β-carotene levels in both groups may have reduced lipid peroxidation in RBC and plasma. Vitamin C increased in both the coriander and anise groups. Therefore, both seeds have antioxidant, hypolipidemic, and antidiabetic effects in diabetics (Rajeshwari et al., 2011) (Table 4.3).

Table 4.3 Pharmacological effects of *Pimpinella anisum*

System	Effect	Preparation	References
Organism	Antibacterial	Aqueous and 50% (v/v) methanol extract	Akhtar et al. (2008)
		Ethanol extract	Ateş and Turgay (2003)
		Essential oil and methanol extract (in combination with Thymus vulgaris)	Al-Bayati (2008)
		Aqueous decoction	Chaudhry and Tariq (2006)
		Essential oil	Shukla and Tripathi (1987)
	Antifungal	Fluid extract	Kosalec et al. (2005)

(Continued)

Table 4.3 (Continued) Pharmacological effects of *Pimpinella anisum*

System	Effect	Preparation	References
		Methanol extract	Yazdani et al. (2009)
	Insecticidal	Essential oil	Park et al. (2006)
		p-Anisaldehyde from aniseed oil	Lee (2004)
	Antiviral	Essential oil	Shukla et al. (1989)
		Lignin-carbohydrate-protein complexes from hot water extract	Lee et al. (2011)
Muscle	Muscle relaxant of tracheal chain	Aqueous extract	Boskabady and Ramazani-Assari (2001)
		Ethanolic extract	Boskabady and Ramazani-Assari (2001)
		Essential oil	Boskabady and Ramazani-Assari (2001)
	Antispasmodic and relaxant of anococcygeus smooth muscle	Hydroalcoholic extract (60% ethanol)	Tirapelli et al. (2007)
Nervous system	Anticonvulsant	Essential oil	Pourgholami et al. (1999)
		Methanol extract of seeds	Heidari and Ayeli (2005)
		Aqueous extract of leaves and stem extract	Abdul-Ghani et al. (1987)
	Analgesic	Essential oil	Twaij et al. (1987)
		Fixed oil	Tas (2009)
	Conditioned place aversion in morphine dependence	Essential oil	Tas et al. (2006)

(*Continued*)

Table 4.3 (Continued) Pharmacological effects of *Pimpinella anisum*

System	Effect	Preparation	References
Gastrointestinal	Antiulcer	Aqueous suspension	Al Mofleh et al. (2007)
	Palliation of nausea	Essential oils of *aniseeds, foeniculum vulgar, Anthemis nobilis, and Mentha piperit*	Gilligan (2005)
	Laxative	Phytotherapic compound of anise and *foeniculum vulgar, Sambucus nigra, Cassia angustifolia*	Picon et al. (2010)
	Increase glucose absorption from the jejunum	Essential oil	Kreydiyyeh et al. (2003)
Endocrine	Antidiabetic hypolipidemic	Seed powder	Rajeshwari et al. (2011)
Renal	Antioxidant	Ethanol extract	Tirapelli et al. (2007)
Immune system		Water extract	Heidari and Ayeli (2005)
		Essential oil	Gilligan (2005)
		Oleoresin	Picon et al. (2010)
		Ethyl acetate fraction of ethanol extract	Abdul-Ghani et al. (1987)
		Anise tea	Boskabady and Ramazani-Assari (2001)
	Increase of β-carotene, vitamins A, C	Seed powder	Yazdani et al. (2009)
Others	Reduction of menopausal hot flashes	Capsules of anise extract	Akhtar et al. (2008)
	Growth promoter of day-old broilers	Essential oil	Park et al. (2006)
	Reduction of pain in dysmenorrhea	Herbal capsule (extracts of anise, celery, saffron)	Akhtar et al. (2008)

4.2.18 Extraction of antioxidants

Al-Wendawi and Gharb (2021) reported antimicrobial, antibiofilm, and antioxidant properties of essential oil extracted from *anise* (PAS) seeds. Using a Clevenger apparatus and a hydro-distillation method, aniseed oil was extracted. Gas chromatographic-mass spectral (GC–MS) analysis identified 25 distinct compounds; the main constituents were anisole (87.99%) and estragole (3.11%), with the remaining constituents accounting for less than 2% of the mixture. In a concentration-dependent manner, the aniseed EOs demonstrated strong antioxidant activity for scavenging DPPH; at 10,000 ppm, the highest antioxidant content (84.9%±0.267%) was observed. Regarding MIC against burn wounds, isolated *Pseudomonas aeruginosa* exhibited antibacterial activity within the 4,000–7,000 ppm range. Aniseed EOs reduced *P. aeruginosa* biofilm formation by 26%–86.41% in the co-treatment assay. Most *P. aeruginosa* candidate isolates had 100% lower pre-developed biofilm inhibition potentials.

Gulcin et al. (2003) stated the screening of the antibacterial along with the antioxidant properties of seed extracts from the anise plant (*Pimpinella anisum* L.). This study looked into the antibacterial and antioxidant properties of anise (*P. anisum*) seed water and ethanol extracts. Both *P. anisum* extracts were tested for antioxidant activity using free radical scavenging, reducing power, superoxide anion radical scavenging, H_2O_2 scavenging, and chelating of a metal. In the linoleic acid system, 20 mg/mL water as well as ethanol extracts inhibited peroxidation by 99.1% and 77.5%, respectively. This was higher than 36.1% for α-tocopherol. Compared to synthetic antioxidants like α-tocopherol, BHT, and BHA, these antioxidant activities were different. *P. anisum* water extract had more antioxidants than ethanol. Ten microbial species were tested for antimicrobial activity by utilizing the disc diffusion approach.

The medicinal plant *Pimpinella anisum* has been utilized in Iranian traditional medicine for a variety of purposes. In an effort to determine the chemical composition and pharmacological characteristics of *P. anisum*, numerous investigations have been conducted on the plant's extracts and essential oil thus far. Reports have indicated that aniseeds possess antimicrobial, antiviral, antifungal, insecticidal, and antioxidant effects. The results also showed that aniseeds have beneficial effects on the digestive system, protect the stomach, and relax muscles. They lower lipid peroxidation and have hypolipidemic and hypoglycemic effects in diabetic patients. In mice, aniseeds also produced conditioned place aversion, decreased morphine dependence, and exhibited anticonvulsant effects. Menopausal hot flashes and dysmenorrhea in women can both benefit from aniseed. The compounds that hold the greatest significance in aniseed essential oil are methyl chavicol, γ-hymachalen, trans-anethole, estragole, and p-anisaldehyde. Given the plant's wide range of pharmacological effects and the paucity of clinical research on it, further clinical trials are advised to assess *Pimpinella anisum*'s beneficial effects in human models and identify the plant's active compounds, which may one day lead to the preparation of new drugs containing these active ingredients.

4.2.19 Rhus coriaria

Rhus coriaria, also called Sicilian sumac, is a small tree or shrub that grows to a height of 1–3 m. There are 9–15 leaflets per imparipinnate leaf. The fruit is a villose, reddish, single-seeded drupe, the flowers are tiny and greenish white, and the inflorescence is an erect, compact panicle (Davis, 1967). According to reports, this plant has the following characteristics: gallotannins, volatile oil, flavonoids, anthocyanin (Güvenc and Koyuncu, 1994), hydrolyzable tannins (Mavlyanov et al., 1997), flavones, involving myricetin, quercetin, and kaempferol (Mehrdad et al., 2009), nitrate and nitrite contents (Özcan and Akbulut, 2007), oil, moisture, protein, fiber, and ash (Özcan and Haciseferogullari, 2004), gallic acid (EI Sissi et al., 1972). Sumac oil primarily consists of palmitic, malic, oleic, stearic, and linoleic acids (Kizil and Turk, 2010). P, K, Br, Si, Cu, Al, S, Cl, Pb, Ti, Ca, Mn, Fe, Zn, Sr, Ba, Mg, Li, Cr, N, and V are among the minerals" observed in plants (Barakat et al., 2003; Özcan, 2004).

Sumac has antimicrobial, antifungal, hepatoprotective, anti-inflammatory/chondroprotective, antioxidant, anti-ischemic, DNA-protective, hypoglycemic, and vasorelaxant properties (Adwan et al., 2010; Gündüz et al., 2010).

From an industrial perspective, *sumac* is used to dye and tan fine leather because it contains coloring matter and tannins. Thus, leaves are exported (Saeed, 1972; Baytop, 1984). *Sumac,* a natural dye, can evenly dye silk, Angora rabbit wool, sheep wool, and goat wool in high light (Kumbasar et al., 2009). It is remarkably effective against wood decay and has high fixation, retention, and fungal resistance qualities (Sen et al., 2009).

4.2.20 Phytochemistry

Many research groups have conducted thorough physical and chemical analyses of *sumac* in light of its significance in the traditional medical system, industry, and as a spice. By using paper chromatography to separate the flavonols myricetin, kaempferol, and quercetin from the tannins found in *sumac,* along with methyl gallate, gallic acid, ellagic acid, and *m*-digallic acid, the leaves of *Rhus coriaria* were extracted. UV spectrophotometric measurements were used to confirm the phenolic components (El Sissi et al., 1966). By using paper chromatography, UV, and IR measurements to assign the structure of an octa- or nona-galloylated glucose to isolated gallotannins from the leaves of *sumac*, more thorough investigations were made possible (El Sissi et al., 1971). Adsorption column chromatography was used to demonstrate the presence of flavonoid glycosides, and an aqueous acetone extract of *sumac* leaves was observed to contain methyl gallate (El Sissi et al., 1972). Liquid chromatography has also been used to isolate the dimeric flavonoids amenthoflavone, agathisflavone, hinokiflavone, and sumaflavone from *sumac* leaves (Van Loo et al., 1988). GC and GC–MS were used to examine the dry fruits' steam volatile components. The primary components identified included aliphatic aldehydes, farnesyl acetone, hexahydrofarnesylacetone, and oxygenated terpenes like carvacrol, α-terpineol, and β-caryophyllene alcohol. Major constituents included terpene hydrocarbons like α-pinene, β-caryophyllene, and cembrene (Figure 4.8) (Brunke et al., 1993). Using normal phase HPLC to analyze the components of the aqueous extract, it was discovered that the extract contained glucose derivatives of penta-, hexa-, hepta-, octa-, nona-, deca-, undeca-, and dodecagallolyl. The ideal method for extracting tannins from *sumac* leaves is to soak them in water at 45°C for 60 minutes without stirring (Zalacain et al., 2003). A monolithic column and reversed phase HPLC were used to separate and quantify flavonol aglycones in *Rhus coriaria* in less than a minute without compromising column efficiency or selectivity (Mehrdad et al., 2009).

More than 20 chemical elements—often referred to as minerals—are essential to human health. Minerals that are necessary for bodily functions must be obtained through diet. Deficiency symptoms could appear if the intake is inadequate (Bender and Mayes, 2003). X-ray fluorescence

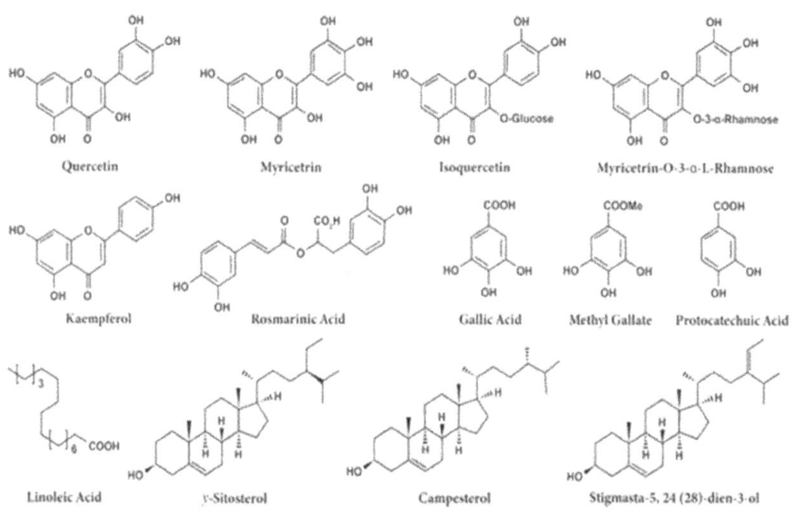

Figure 4.8 Structure of phytochemical compounds extracted from *Rhus Coriria*.

analysis of pallets of *Rhus coriaria* from Jordan revealed the existence of Ti, Fe, Cu, Cl, Al, P, Mn, K, Zn, Ca, Br, Si, S, and Sr in varying amounts (Barakat et al., 2003). The sumac fruits were then dried, ground, and burned in an oven with pure HNO_3. In addition to Ba, Mg, Li, Cr, Pb, N, and V, numerous other minerals have also been identified using inductively coupled plasma atomic emission spectrometry (Özcan, 2004). Sumac fruits were found to contain high concentrations of potassium, calcium, magnesium, and phosphorus (Kizil and Turk, 2010; Özcan and Akbulut, 2007; Özcan and Haciseferogullari, 2004; Barakat et al., 2003).

Cold ether extraction was used to extract sumac oil. Gas chromatography shows that sumac contains mostly oleic (C18:1), linoleic (C18:2), and palmitic (C16:0) fatty acids. The percentage of polyunsaturated fatty acids (18:2 + 18:3) was 34.84%–37.36% (Dogan and Akgul, 2005). The sumac fruit was extracted with petroleum ether and analyzed using gas chromatography-mass spectroscopy. Several fatty acids were found in *Rhus coriaria,* including arachidic, stearic, tetradecanoic, behenic, linoleic, malic, 7-hexadecenoic, azelaic, palmitic, ethyl octadecenoate, oleic, and elaidic acids. Sumac fruits contain mainly oleic, linoleic, and palmitic acids. Moreover, fruits from Turkey's Mersin region have 1.5% ash and 2.3% proteins (Kizil and Turk, 2010). Using a hot distilled water extract of sumac fruits, the American Public Health Association's diazotization and phenoldisulphonic acid method was used to determine the components of nitrate and nitrite. According to Özcan and Akbulut (2007), the concentrations of nitrite and nitrate were determined to be 162.28 and 36.60 mg/kg, respectively. Figure 4.8 presents the significant compounds identified from R. coriaria.

The length, weight, geometric mean diameter, volume, sphericity, thickness, and moisture content level of *sumac* fruits at 4.79% were determined to be 0.018 g, 4.72 mm, 19.49 mm, 3.64 mm, 0.773 mm, and 2.64 mm, respectively. Porosity, bulk density, static friction, projected area, terminal velocity, and 304.25 kg/m^3, 3.52 m/s, and 0.164 cm^3 of *sumac* fruits were evaluated at the same moisture content level (Özcan and Haciseferogullari, 2004).

4.2.21 Use in traditional medicine

Traditional uses of *sumac* include treating diarrhea "(Saeed, 1972; Usmanghani et al., 1997; Sezik et al., 2001; Said et al., 2002; Lev and Amar, 2002, 2008; Lev, 2007), hemorrhoids (Saeed, 1972; Lev, 2002), ulcers (Tuzlacı and Aymaz, 2001), liver disease (Said et al., 2002), bites of animal, pain (Lev, 2002), diuresis, dysentery, hematemesis, hemorrhage, ophthalmia, hemoptysis, leucorrhea, stomach tonic, and conjunctivitis (Saeed, 1972; Usmanghani et al., 1997). Additionally, *sumac* has been utilized by traditional medical practitioners as an abortifacient, to treat sore throats, and to lower cholesterol (Lev and Ammar, 2002; Lev, 2007; Lev and Amar, 2008). According to other reports", it is also used as an antimicrobial and to heal wounds (Rayne and Mazza, 2007).

The traditional medical system has employed various plant parts in various preparations. Its bark powder works well for teeth cleaning. When viral eye infections are first developing, bark infusion is helpful. Bark bruised in the water is applied to the forehead for epistaxis first aid (Usmanghani et al., 1997). Consuming powdered fruits on boiled eggs treats diarrhea (Sezik et al., 2001). For the treatment of diarrhea, disease of the liver, and UTI, a fruit decoction (150 cc) is prepared and taken orally. The decoction is taken thrice daily until the condition improves (Said et al., 2002).

4.3 *Pharmacological properties*

4.3.1 *Antioxidant property*

Peanut oil's autoxidation was stopped by *sumac*. Peroxide value was measured in order to determine antioxidant activity (Özcan, 2003). Lipid peroxidation is the cause of many diseases, including emphysema, mutagenesis, heart disease, stroke, and aging. Ethyl acetate from the "methanol: water (80:20)" extract had strong antiradical activity on DPPH. It also inhibited Fe^{2+}-induced linoleic acid peroxidation (Bozan et al., 2003). Kosar et al. (2007) found that anthocyanin and hydrolyzable tannin derivatives inhibit lipid peroxidation and DPPH radical scavenging in the HPLC-MS system. Free radicals are generally thought to cause harm to proteins, carbohydrates, lipids, and nucleic acids. *Rhus coriaria* fruit methanol extracts prevented the hydroxyl radical generated by the Fe^{3+}/ascorbate/EDTA/H_2O_2 system from degrading deoxyribose. The same extract demonstrated antioxidant activity in vitro by scavenging superoxide radicals generated by the xanthine/xanthine oxidase method. *Rhus coriaria* also appears to inhibit the rat liver homogenates' production of lipid peroxidase by Fe^{2+}-ascorbate (Candan and Sökmen, 2004). ROS, nitric oxide, and inflammatory cytokines gradually deteriorate the articular cartilaginous tissue, which may result in osteoarthritis. *Sumac* leaf lyophilized hydroalcoholic crude extract decreased oxidative stress, chondrodegeneration, and inflammation. It enhanced the synthesis of glycosaminoglycans and decreased the production of prostaglandins E_2, ROS, and nitric oxide brought on by interleukin-1β. These actions may be partially attributed to the polyphenolic active constituents found in *sumac* (Panico et al., 2009). A major factor in the beginning as well as progression of several liver diseases is oxidative stress. In isolated rat hepatocytes, aqueous extract of the *Rhus coriaria* fruits inhibited the oxidative stress toxicity induced by cumene hydroperoxide. It significantly shielded the hepatocytes from oxidative damage to the lysosomal membrane, lipid peroxidation, cellular proteolysis, glutathione depletion, and a decrease in mitochondrial membrane potential. One of the main components of plants, gallic acid, exhibited a comparable level of defense (Pourahmad et al., 2010).

4.3.2 Antibacterial property

Rhus coriaria leaf methanol extract inhibited several Gram-positive and Gram-negative bacteria strains. MICs and MBCs have been determined by plating 0.01 mL samples from clear 1 mL tubes on the agar plates and performing a Mueller–Hinton broth microdilution test in microliter plates. *Susan* had the highest activity against the *E. coli* ATCC 30213 at 156 mg/L among Gram-negative strains. Iauk et al. (1998) found that it inhibited *Bacillus subtilis* at 78 mg/L and *Staphylococcus aureus* clinical and standard isolates at 312 and 156 mg/L. *Rhus coriaria* fruit extracts in chloroform, ethyl acetate, and n-butanol inhibited *Branhamella catarrhalis*. The ethyl acetate extract had the highest inhibition zone and activity, with a MIC of 125 μg/mL, while the aqueous part showed no activity (Sökmen et al., 1999). Fruit chloroform extract demonstrated potent antibacterial activity against *Bacillus megaterium*, *E. coli*, *Bacillus cereus*, *Bacillus bravis*, *Listeria monocytogenes*, *Enterobacter aerogenes*, *P. aeruginosa*, and *Micrococcus luteus* in a disk diffusion method, with an inhibition zone of 35–51 mm (Dığrak et al., 2001). Fifteen *Lactobacillus* species, including *L. plantarum* C32, *Lactobacillus plantarum* C27, *L. plantarum* P33, *L. plantarum* C39, *L. plantarum* P53, *L. plantarum* P51, *L. fermentum* P37, *L. fermentum* C47, *L. fermentum* P46, *L. fermentum* P38, *L. coryniformis* subsp. *torquens* C33, *L. fermentum* P54, *L. acidophilus* P49, *L. animalis* C26, and *Lactobacillus* sp. C4 were inhibited by fruits' methanol crude extract, which has been resolved in absolute alcohol at a ratio of 1:5 w/v extract/absolute alcohol. A zone of inhibition measuring between 13 and 22 mm was observed in the plant extract (Sağdic et al., 2003). After the sumac fruits were dried, methanol extract was made. Using the agar well-diffusion approach, the observed diameter of the inhibition zone against *Bordetella bronchiseptica*, *Klebsiella pneumoniae*, *Bacillus pumilus*, and *Staphylococcus epidermidis* was greater than 15 mm. Eighteen months later, the extract exhibited the same activity (Bonjar, 2004). The ethanol extract of *Rhus coriaria* fruits did not show any anti-yeast effect, but it was effective against *Mycobacterium smegmatis* and *P. vulgaris*, showing inhibition zones of 25 and 38 mm, correspondingly, which included a disc with a diameter of 6 mm (Dulger and Gonuz, 2004). Gram-positive bacteria are more susceptible than Gram-negative bacteria to the inhibitory effect of an alcoholic sumac extract. The cup method and MIC were used. By tracking each test organism's growth on the nutrient agar comprising sumac extract at many incremental concentrations ranging from 100 to 5,000 mg/L, the MIC for each was determined. Only 500 mg/L of the sumac was sufficient to cause death in *Bacillus* species (*Bacillus megaterium, B. thuringiensis, B. cereus*, and *B. subtilis*), *Staphylococcus aureus* (1,000 mg/L), and *Listeria monocytogenes* (1,500 mg/L). *P. vulgaris, C. freundii, E. coli*, and *H. alvei* O157:H7 were among the Gram-negative bacteria that could withstand doses of up to 1,000, 1,500, 2,000, and 2,500 mg/mL, correspondingly. *Salmonella enteritidis* and *Escherichia coli* type I showed greater resistance, surviving up to 3,000 mg/L. The brown-red extract of ripe sumac outperformed the greenish extract of unripe sumac (Nasar-Abbas et al., 2004). Clinical bacterial strains, including *EHEC, Pseudomonas aeruginosa, P. aeruginosa, P. vulgaris, Klebsiella pneumoniae*, and methicillin-resistant *Staphylococcus aureus*, were inhibited by ethanol and methanol extracts of sumac fruits; however, the water extract had no effect on *P. aeruginosa, P. vulgaris*, or *Klebsiella pneumoniae*. Techniques for well diffusion and microdilution were used. The reference strain was *B. subtilis*, and the outcomes were contrasted with those of the antibiotic tetracycline. According to Abu-Shanab et al. (2005), the MIC of ethanol extract was 0.156 mg/mL for *B. subtilis* and 1.25 mg/mL for *EHEC, MRSA, Proteus vulgaris*, and *P. aeruginosa*. *Rhus coriaria* hydroalcoholic extract was prepared by percolating 80% v/v aqueous alcohol. It was discovered to be effective against *Shigella flexneri* and *Salmonella typhi*, with MICs of 0.20% and MBCs of 0.80% and 0.40%, respectively, for each bacterium (Fazeli et al., 2007). An 80% ethanol extract of sumac seeds was found to interact synergistically with antimicrobial agents, including enrofloxacin, sodium sulfadimethoxine, penicillin G, cephalexin, and oxytetracycline HCl. According to Adwan et al. (2010), this combination may help combat *P. aeruginosa* that is developing drug resistance. The *Salmonella typhimurium* nalidixic acid-resistant strain, which has been inoculated into tomatoes, was significantly reduced by sumac water extract, which can be used as a natural substitute for chlorine in bacterial inhibition. According to Gündüz et al. (2010), the plant extract may help improve the tomatoes' microbiological safety without sacrificing their quality attributes.

4.3.3 Non-mutagenic property

For mutagenic studies, *Salmonella typhimurium* strains TAg8, TA97a, TA102, and TA100 were employed. *Sumac* has no base pair substitution mutagenic activity, according to a plate incorporation test (Barakat et al., 2003).

4.3.4 Hypoglycemic property

At 100 µg/mL, the methanol *Rhus coriaria* fruits' extract showed 48.3% inhibition of α-amylase, while the ethyl acetate and *n*-hexane extracts showed 91.9% and 44% inhibition, respectively. Therefore, sumac, particularly the ethyl acetate extract, with an IC50 value of 28.7 µg/mL, might help diabetic patients' glucose tolerance (Giancarlo et al., 2006).

Rats with alloxan-induced diabetes demonstrated notable hypoglycemic effects when exposed to *Rhus coriaria* fruit ethanol extract. Five hours following the administration of a single dose, there was a 24% decline in blood glucose (at a 400 mg/kg dose). In the long-term test, there was a 26% drop in blood glucose levels on day 21 compared to the diabetic control group. In the oral glucose tolerance test, a major drop in blood glucose was noted 60 minutes after the carbohydrate solution was administered (at 200 and 400 mg/kg doses). Ethanol extract also markedly decreased the intestinal maltase and sucrase activities (44.33% and 25.38%, respectively) (Mohammadi et al., 2010).

4.3.5 *Anti-migratory property*

After grinding dry sumac, tannins were then extracted. Tannins were applied to cultured rat carotid vascular smooth muscle cells. The vascular smooth muscle cells' migratory activity in response to platelet-derived growth factor-BB was measured using the transmembrane migration assay. *Sumac* has strong anti-migratory activity and has been shown to significantly reduce vascular smooth cell migration by 62%. It may also have atheroprotective properties (Zargham and Zargham, 2008).

4.3.6 *Anti-ischemic property*

An extract from *sumac* leaves showed heart-protective properties. Thoracic aorta and isolated rabbit heart preparations were used to study this property. The cardiovascular protective effect has been attributed to various factors, including the scavenging of free radicals and ROS, inhibition of TNF-α, activation of the cyclooxygenase pathway, and endothelial nitric oxide synthase activation. According to Baretta et al. (2009), the hydrolyzable gallotanins found in *sumac* leaves are primarily responsible for their anti-ischemic effects.

4.3.7 *DNA protective effect*

Sumac significantly decreased DNA migration following a 30% H_2O_2 treatment of the cells. Additionally, there was a significant 36% and 52% decrease in DNA migration caused by endogenous production of oxidized pyrimidines and purines, respectively. The liver and lymphocytes showed the greatest reduction in purine and pyrimidine oxidation-related damage. Drinking water supplementation evidently reduced comet formation due to oxidized DNA bases. The sensitivity of cells to anti-benzo[*a*]pyrene-7,8-dihydro-diol-9,10-epoxide (BPDE) was also altered, as comet formation induced by BPDE was significantly decreased by 69%. According to Chakraborty et al. (2009), there was a noticeable increase in the overall GST activity in the plasma along with the two isoenzymes GST-π and GST-α by 40%, 26%, and 52%, respectively. This indicates that *sumac* may offer protection against carcinogens that are genotoxic and are broken down by these enzymes.

4.3.8 Anti-fungal property

Rhus coriaria leaf methanol extract was susceptible to *Candidia albicans* at 625 mg/L MIC (Iauk et al., 1998). To calculate the total fungal counts in spice samples, the dilution method was applied. Three distinct nutritive media, including Czapek Dox agar, Coca-Cola Rose, and potato dextrose agar, were selected. *Sumac* demonstrated an antifungal effect by showing a very low affinity to be contaminated with mold (Hashem and Alamri, 2010).

Over time, numerous traditional uses of *sumac* have been validated by scientists as important natural therapies for a variety of ailments. From a medicinal perspective, *sumac* has qualities that are antimicrobial, antifungal, hepatoprotective, anti-inflammatory/chondroprotective, antioxidant, vasorelaxant, anti-ischemic, hypoglycemic, anti-migratory, inhibit XO, protect DNA, and are not mutagenic. It is a highly regarded condiment that is used as a major souring agent from a nutritional standpoint. Blended with freshly cut onion, it's served as an appetizer. Some Turkish fast food, particularly döner kebab, has a hint of *sumac* powder. Additionally, kebabs, chicken, or fish are rubbed with it (Baytop, 1984). Minerals found include Pb, Mg, Li, Cr, Ba, N, and V, as well as Ti, Fe, Cu, Cl, Al, K, P, Ca, Mn, Si, Zn, S, Br, and Sr. Iron is an essential component of the human body that aids in the transfer of oxygen and electrons. It is necessary for the production of hemoglobin (Kaya and Incekara, 2000; Dalziel, 1936). The primary component of bone, calcium aids in the growth of teeth (Anke et al., 1984). Due to their numerous biological functions, including participation in enzymatic and redox reactions, zinc and copper are essential components of our diets (McLaughlin et al., 1999). Both human and animal research has shown that adequate intake of minerals like magnesium, calcium, zinc, potassium, copper, and manganese can lower individual cardiovascular disease risk. Also important are Mg, P, and Fe levels. While not necessary, lead, bromine, and barium aid biological processes (Macrae et al., 1993a). Lithium, which treats manic-depressive disorders, is another important mineral (Macrae et al., 1993b).

Rhus coriaria plays a significant role in the leather industry as a tanning agent. Goatskin processed with *sumac* yielded the best quality bookbinding leather, and Morocco leather was originally tanned with *sumac*. There are several advantages to using plants as coloring agents in the textile industry, particularly when dyeing textile materials made of protein and preventing wood decay.

4.3.9 Extraction of antioxidants

Rhus coriaria L., commonly referred to as sumac (*Rhus coriaria*), has several pharmacological reputations in Western Asia, Southeastern Anatolia, the Mediterranean region, and Iran. These reputations make the extracts from the plant important for drug development. The objective of this study was to analyze the ethanolic extract of Iranian sumac (*Rhus coriaria* L.) fruit for its antimicrobial and antioxidant properties. Using the microdilution method, the antimicrobial activity of the fruit ethanolic extract of *R. coriaria* L. was evaluated against bacterial strains obtained from food, including both gram-positive and gram-negative bacteria. Additionally, the ethanolic extract's antioxidant properties were examined, including its capacity to scavenge DPPH radicals. The GC/MS method was also used to investigate the chemical compositions of the essential oil (EO) extracted from sumac fruit. For every tested species of bacteria, the extract demonstrated potent antimicrobial activity with concentration dependence and a wide antibacterial spectrum. The most sensitive Gram-positive and Gram-negative bacteria were determined to be *S. aureus* and *Salmonella enterica*, with a MIC of less than 0.78%, respectively. The ethanolic sucrose extract exhibited a strong antioxidant effect. In every concentration tested, the sumac ethanolic extract's antioxidant capacity outperformed BHT's. The fruit's essential oil was observed to consist of 11 various constituents. The primary constituents observed in the EO were diethyl ester (22.01%), butanedioic acid (39.7%), and malate. This research raises the prospect of employing *R. coriaria* L. fruit as a novel source of naturally occurring antioxidants and antimicrobials for the food and pharmaceutical sectors (Mahdavi et al., 2018).

The antioxidant activity and the phenolic composition of *sumac* (*Rhus coriaria* L.) extracts were reported by Kosar et al. (2007). The study examined the antioxidant activity along with the phenolic compounds present in *sumac* extracts. After *sumac* was extracted in methanol, it underwent solvent-solvent partitioning, producing two fractions: aqueous and ethyl acetate. A Sephadex LH-20 column was used to further fractionate the methanol extract. DPPH radical scavenging and ferric thiocyanate were used to screen extracts and fractions for antioxidant activity. The active fraction(s)' phenolic composition was ascertained using HPLC–MS systems. Rich in hydrolyzable tannins along with anthocyanins, the fractions showed strong antioxidant activity. The anthocyanin fraction included peonidin, cyanidin, petunidin, pelargonidin, and delphinidin glucosides and coumarates, but gallic acid constituted the majority of the phenolic acid in the extracts. There was a high amount of pentagalloyl glucose in the hydrolyzable tannin fraction. The anthocyanin-rich fraction had an effective scavenging concentration (EC_{50}) of 5.33 lg/mL, while the ethyl acetate and tannin fractions had an EC_{50} of 0.70 lg/mL. When compared to synthetic antioxidants, the same extracts and fractions demonstrated a moderately inhibiting effect on lipid peroxidation. The results show that *sumac* is a useful natural antioxidant.

Sumac has a lot of potential for further study. For centuries, traditional medicine has utilized it as a stomach tonic and to treat diarrhea "(Usmanghani et al., 1997; Lev, 2007; Lev and Amar, 2008). Tannins and flavonoids, which are known to have spasmolytic and antidiarrheal properties, were found through phytochemical screening (Di Carlo et al., 1993; Palombo, 2006). In vivo antidiarrheal and in vitro spasmolytic activities must be carried out because the plant is also known to have antibacterial activity, which may provide additional benefit for its utilization in the management of various types of diarrhea, including infectious diarrhea. *Rhus coriaria* has demonstrated hepatoprotective and anti-ischemic qualities in vitro; however, these findings need to be verified in vivo on animals. According to Katzung and Chatterjee (2003), *Rhus coriaria* is also said to contain nitrates along with nitrites, both of which relax vascular smooth muscles and cause vasodilation. Vasodilators lower peripheral vascular resistance, which reduces hypertension (Benowitz, 2003). It is known that flavonoids have the ability to block the Ca^{2+} channel (Revuelta et al., 1997; Gilani et al., 2007). Ca^{2+} channel blockers both vasodilate blood vessels and have a negative inotropic effect on hypertension (Seth, 2009). Therefore, it is recommended that in vivo as well as in vitro experiments be carried out to assess the possibility that the plant has antihypertensive properties. Because there are insufficient clinical data, it is unclear to what extent the findings regarding pharmacological activities have potential clinical relevance. Therefore, scientists still have a large research gap that they can fill in the future to benefit humanity.

4.3.10 Thermal and non-thermal extraction methods for obtaining antioxidants from medicinal fruits

Thermal extraction techniques involve using heat to extract antioxidants from plant materials. These methods often rely on the application of high temperatures to break down cellular structures and release valuable compounds. For plants like *Foeniculum vulgare* (fennel), *Carum carvi* (caraway), *Pimpinella anisum* (anise), and *Rhus coriaria* (sumac), thermal extraction methods such as steam distillation, hot solvent extraction, and microwave-assisted extraction are commonly employed.

Steam Distillation: This technique utilizes steam to extract essential oils and antioxidants. In *Foeniculum vulgare*, steam distillation is effective in obtaining fennel oil rich in antioxidants like flavonoids and phenolic compounds. The steam passes through the plant material, causing the essential oils and antioxidants to evaporate and then condense into a liquid, which is collected and separated (Topal et al., 2008).

Hot Solvent Extraction: In this method, plant materials are soaked in a hot solvent, such as ethanol or hexane. The heat helps to dissolve the antioxidants from the plant matrix into the solvent. For *Carum carvi*, this technique extracts caraway oil, which contains carvacrol and other antioxidants. Similarly, *Pimpinella anisum* can be processed to obtain anise oil rich in anethole and other beneficial compounds (Yeasmen and Islam, 2015).

Microwave-Assisted Extraction: This modern approach uses microwave radiation to heat the solvent and plant material simultaneously. The microwaves create rapid heating, which enhances the extraction of antioxidants. In *Rhus coriaria*, this technique can be used to extract sumac oil or extracts rich in polyphenols and tannins. The efficiency of microwave-assisted extraction makes it a favorable method for obtaining high yields of antioxidants (Zheng et al., 2011)

4.4 Non-thermal extraction techniques

Non-thermal extraction techniques do not involve the use of heat and instead rely on physical or chemical processes to extract antioxidants. These methods are often gentler on the plant material and can preserve heat-sensitive compounds.

Cold Pressing: This mechanical method involves pressing the plant material without the application of heat. For *Foeniculum vulgare* and *Carum carvi*, cold pressing can be used to extract essential oils and antioxidants while preserving their chemical integrity. This method is particularly useful for obtaining high-quality oils with minimal degradation of sensitive antioxidants (Cakaloglu et al., 2018).

Supercritical Fluid Extraction (SFE): This technique uses supercritical fluids, like CO_2, to extract antioxidants. The supercritical fluid acts as a solvent with both liquid and gas-like properties, allowing it to penetrate the plant material and dissolve antioxidants effectively. For *Pimpinella anisum*, SFE can efficiently extract anethole and other antioxidants with high purity and yield (Veggi et al., 2011).

Ultrasound-Assisted Extraction: This method utilizes ultrasonic waves to form cavitation bubbles in the solvent, which helps to disrupt the plant cell walls and release antioxidants. In the case of *Rhus coriaria*, ultrasound-assisted extraction can be used to obtain sumac extracts rich in flavonoids and polyphenols. This technique is known for its efficiency and ability to extract antioxidants at lower temperatures compared to traditional methods (Pingret et al., 2013).

Enzyme-Assisted Extraction: This technique involves the utilization of specific enzymes to break down the plant material's cell walls, facilitating the release of antioxidants. For *Carum carvi* and *Pimpinella anisum*, enzyme-assisted extraction can help obtain essential oils and extracts rich in antioxidants by selectively targeting and breaking down cell components (Puri et al., 2012).

Each of these extraction techniques offers unique advantages and can be selected based on the specific requirements of the plant material and the desired antioxidant profile.

4.5 Conclusion

In conclusion, the extraction of antioxidants from medicinal fruits native to Iraq holds significant potential for both therapeutic applications and as natural preservatives in the food industry. This chapter has highlighted the importance and efficacy of extracting antioxidants from four notable medicinal fruits: *Foeniculum vulgare* (fennel), *Carum carvi* (caraway), *Pimpinella anisum* (anise), and *Rhus coriaria* (sumac). *Foeniculum vulgare*, commonly known as fennel, has demonstrated substantial antioxidant properties due to its high content of phenolic compounds and flavonoids. These antioxidants contribute to the reduction of oxidative stress and provide various health benefits, including anti-inflammatory and antimicrobial effects. The extraction methods for fennel have been optimized to preserve its bioactive compounds, making it a valuable resource in both medicinal and culinary contexts.

Carum carvi, or caraway, is another medicinal fruit with notable antioxidant capacity. The essential oils and extracts from caraway seeds are rich in phenolic acids and flavonoids, which play a crucial role in neutralizing free radicals and preventing cellular damage. The methods employed for extracting these antioxidants have been refined to enhance yield and efficacy, ensuring that the therapeutic potential of caraway is fully realized.

Pimpinella anisum, known as anise, has also been extensively studied for its antioxidant properties. Anise seeds contain anethole and other phenolic compounds that exhibit strong antioxidant activity. The extraction processes for anise have focused on maximizing the retention of these compounds, thereby enhancing its potential for promoting health and preventing diseases associated with oxidative stress.

Lastly, *Rhus coriaria*, commonly referred to as sumac, is renowned for its potent antioxidant properties. Sumac fruits are rich in tannins and anthocyanins, both of which contribute to their high antioxidant activity. The extraction techniques for sumac have been developed to ensure the efficient isolation of these compounds, making sumac a valuable addition to both medicinal formulations and natural food preservative.

Overall, the extraction of antioxidants from these medicinal fruits not only underscores their potential in both traditional and modern medicine but also highlights the importance of optimizing extraction methods to retain the maximum bioactive compounds. The continued exploration and development of these techniques will undoubtedly contribute to the advancement of natural antioxidant sources, ultimately promoting better health and well-being.

References

Abdul-Ghani, A. S., El-Lati, S. G., Sacaan, A. I., Suleiman, M. S., & Amin, R. M. (1987). Anticonvulsant effects of some Arab medicinal plants. *International Journal of Crude Drug Research*, 25(1), 39–43.

Abed, K. F. (2007). Antimicrobial activity of essential oils of some medicinal plants from Saudi Arabia. *Saudi Journal of Biological Sciences*, 14, 53–60.

Abu-Shanab, B., Adwan, G., Abu-Safiya, D., Adwan, K., & Abu-Shanab, M. (2005). Antibacterial activity of Rhus coriaria L. extracts growing in Palestine. *Journal of Islamic University of Gaza (Natural Sciences Series)*, 13, 147–153.

Aćimović, M., Filipović, V., Urosević, M., Oljača, S., Dolijanović, Ž., Dojčinović, N., & Gršić, V. (2016). Influence of temperature, precipitation and insolation on the development rate of annual caraway (Carum carvi var. annua). *Seventh International Scientific Symposium "Agrosym 2016"*, Jahorina, Bosnia and Herzegovina.

Adwan, G., Abu-Shanab, B., & Adwan, K. (2010). Antibacterial activities of some plant extracts alone and in combination with different antimicrobials against multidrug-resistant Pseudomonas aeruginosa strains. *Asian Pacific Journal of Tropical Medicine*, 3, 266–269.

Agarwal, R., Gupta, S. K., Agarwal, S. S., Srivastava, S., & Saxena, R. (2008). Oculohypotensive effects of Foeniculum vulgare in experi mental models of glaucoma. *Indian Journal of Physiology and Pharmacology*, 52, 77–83.

Akhtar, A., Deshmukh, A. A., Bhonsle, A. V., Kshirsagar, P. M., & Kolekar, M. A. (2008). In vitro antibacterial activity of Pimpinella anisum fruit extracts against some pathogenic bacteria. *Veterinary World*, 1(9), 272.

Al-Bataina, B. A., Maslat, O. A., & Al-Kofahi, M. M. (2003). Element analysis biological studies on ten oriental spices using XRF and Ames test. *Journal of Trace Elements in Medicine and Biology*, 17, 85–90.

Al-Bayati, F. A. (2008). Synergistic antibacterial activity between Thymus vulgaris and Pimpinella anisum essential oils and methanol extracts. *Journal of Ethnopharmacology*, 116(3), 403–406.

Albert-Puleo, M. (1980). Fennel and anise as estrogen agents. *Journal of Ethnopharmacology*, 2, 337–344.

Allameh, A., Dadkhah, A., Rahbarizadeh, F., Ashrafi-Helan, J., & Fatemi, F. (2013). Effect of dietary caraway essential oils on expression of β-catenin during 1,2-dimethylhydrazine-induced colonic carcinogenesis. *Journal of Natural Medicines*, 67, 690–697.

AlMofleh, I. A., Alhaider, A. A., Mossa, J. S., Al-Soohaibani, M. O., & Rafatullah, S. (2007). Aqueous suspension of anise "Pimpinella anisum" protects rats against chemically induced gastric ulcers. *World Journal of Gastroenterology: WJG*, 13(7), 1112.

Al-Wendawi, S. A., & Gharb, L. A. (2021). Antioxidant, antibacterial and antibiofilm potentials of anise (Pimpinella anisum) seeds extracted essential oils. *Iraqi Journal of Agricultural Sciences*, 52(2), 348–358.

Aly, A., Maraei, R., Rezk, A., & Diab, A. (2023). Phytochemical constitutes and biological activities of essential oil extracted from irradiated caraway seeds (Carum carvi L.). *International Journal of Radiation Biology*, 99(2), 318–328.

Amin, G. R. (2005). Popular medicinal plants of Iran, vice chancellorship of research. Tehran University of Medical Science Press, Tehran, Iran.

Anke, M., Groppel, B., & Kronemann, H. (1984). Significance of newer essential trace elements (like Si, Ni, As, Li, V,…) for the nutrition of man and animals. In: Bratter, P. and P. Schramel. Eds., *Trace Element-Analytical Chemistry in Medicine and Biology* (Vol. 3, pp. 421–464). Berlin: Walter de Gruyter & Co.

Ateş, D. A., & Turgay, Ö. (2003). Antimicrobial activities of various medicinal and commercial plant extracts. *Turkish Journal of Biology*, 27(3), 157–162.

Baananou, S., Bagdonaite, E., Marongiu, B., Piras, A., Porcedda, S., Falconieri, D., & Boughattas, N. (2013). Extraction of the volatile oil from Carum carvi of Tunisia and Lithuania by supercritical carbon dioxide: chemical composition and antiulcerogenic activity. *Natural Product Research*, 27(22), 2132–2136.

Baretta, G., Rossoni, G., Santagati, N. A., & Facino, R. M. (2009). Anti-ischemic activity and endothelium dependent vasorelaxant effect of hydrolysable tannins from the leaves of Rhus coriaria (Sumac) in isolated rabbit heart and thoracic aorta. *Planta Medica*, 75, 1482–1488.

Baytop, T. (1984). *Therapy with Medicinal Plants in Turkey (Past and Present)* (1st Ed., 3255p), Istanbul (Turkey): Istanbul University, (In Turkish).

Bell, G. (2001). Neutral macroecology. *Science*, 293(5539), 2413–2418.

Bender, D. A, & Mayes, P. A. (2003). Vitamins & minerals. In: Murray, R. K., D. K. Granner, P. A. Mayes, and V. W. Rodwell. Eds., Harper's Illustrated Biochemistry (26th Ed., pp. 481–497). NewYork: Lange Medical Books/ McGraw-Hill Medical Publishing Division.

Benowitz, N. L. (2003). Antihypertensive agents. In: Katzung, B. G., S. B. Masters, and A. J. Trevor, Eds., *Basic and Clinical Pharmacology* (9th Ed., pp. 172–180). New York, USA: McGraw-Hill Companies.

Berman, S. L., Wicks, A. C., Kotha, S., & Jones, T. M. (1999). Does stakeholder orientation matter? The relationship between stakeholder management models and firm financial performance. *Academy of Management Journal*, 42(5), 488–506.

Besharati-Seidani, A., Jabbari, A., & Yamini, Y. (2005). Headspace solvent microextraction: a very rapid method for identification of volatile components of Iranian Pimpinella anisum seed. *Analytica Chimica Acta*, 530(1), 155–161.

Bonjar, S. (2004). Evaluation of antibacterial properties of some medicinal plants used in Iran. *Journal of Ethnopharmacology*, 94, 301–305.

Boon, T., Coulie, P. G., Eynde, B. J. V. D., & Bruggen, P. V. D. (2006). Human T cell responses against melanoma. *Annual Review of Immunology*, 24(1), 175–208.

Boskabady, M. H., & Ramazani-Assari, M. (2001). Relaxant effect of Pimpinella anisum on isolated guinea pig tracheal chains and its possible mechanism(s). *Journal of Ethnopharmacology*, 74(1), 83–88.

Bozan, B., Kosar, M., Tunalier, Z., Ozturk, N., & Baser, K. H. C. (2003). Antioxidant and free radical scavenging activities of Rhus coriaria and Cinnamomum cassia extracts. *Acta Alimentaria*, 32, 53–61.

Brunke, E. J., Hammerschmidt, F. J., Schmaus, G., & Akgul, A. (1993). The essential oil of Rhus coriaria L. fruits. *Flavour and Fragrance Journal*, 8, 209–214.

Burkhardt, G., Reichling, J., Martin, R., & Becker, H. (1986). Terpene hydrocarbons in Pimpinella anisum L. *Pharmaceutisch Weekblad*, 8, 190–193.

Cakaloglu, B., Ozyurt, V. H., & Otles, S. (2018). Cold press in oil extraction. A review. *Ukrainian Food Journal*, 7(4), 640–654.

Campbell-Hunt, C. (2000). What have we learned about generic competitive strategy? A meta-analysis. *Strategic Management Journal*, 21(2), 127–154.

Candan, F., & Sökmen, A. (2004). Effects of Rhus coriaria L. (Anacardiaceae) on lipid peroxidation and free radical scavenging activity. *Phytotherapy Research*, 18, 84–86.

Cantore, P. L., Iacobelli, N. S., Marco, A. D., Capasso, F., & Senatore, F. (2004). Antibacterial activity of Coriandrum sativum L. and Foenic ulum vulgare Miller Var. vulgare (Miller). essential oils. *Journal of Agricultural and Food Chemistry*, 52, 7862–7866.

Casirola, D. M., & Ferraris, R. P. (2006). α-Glucosidase inhibitors prevent diet-induced increases in intestinal sugar transport in diabetic mice. *Metabolism*, 55(6), 832–841.

Chakraborty, A., Ferk, F., Simic, T., Branter, A., Dusinska, M., Kundi, M. C., Nersesvan, A., & Knasmuller, S. (2009). DNA-protective effects of sumach (Rhus coriaria L.), a common spice: results of human and animal studies. *Mutation Research*, 661, 10–17.

Charles, D. J. Ed. (2012). Caraway. In: *Antioxidant Properties of Spices, Herbs and Other* Sources (pp. 199–206). New York: Springer.

Chaudhry, N. M., & Tariq, P. (2006). Bactericidal activity of black pepper, bay leaf, aniseed and coriander against oral isolates. *Pakistan Journal of Pharmaceutical Sciences*, 19(3), 214–218.

Cheplick, S., Kwon, Y. I., Bhowmik, P., & Shetty, K. (2010). Phenolic-linked variation in strawberry cultivars for potential dietary management of hyperglycemia and related complications of hypertension. *Bioresource Technology*, 101(1), 404–413.

Chiasson, J. L., Josse, R. G., Gomis, R., Hanefeld, M., Karasik, A., & Laakso, M. (2002). Acarbose for prevention of type 2 diabetes mellitus: the STOP-NIDDM randomised trial. *The Lancet*, 359(9323), 2072–2077.

Choi, E. M., & Hwang, J. K. (2004). Anti-inflammatory, analgesic and antioxidant activities of the fruit of Foeniculum vulgare. *Fitoterapia*, 75(2004), 557–565.

Choudhary, N., Khajuria, V., Gillani, Z. H., Tandon, V. R., & Arora, E. (2014). Effect of Carum carvi, a herbal bioenhancer on pharmacokinetics of antitubercular drugs: a study in healthy human volunteers. *Perspectives in Clinical Research*, 5(2), 80–84.

Collier, A., Small, M., Wilson, R., Bradley, H., & Thomson, J. A. (1990). Free radical activity in type 2 diabetes. *Diabetic Medicine*, 7(1), 27–30.

Dadalioglu, I., & Evrendilek, G. A. (2004). Chemical compositions and antibacterial effects of essential oils of Turkish oregano (Origanum minutiflorum), bay laurel (Laurus nobilis), Spanish lavender (Lav andula stoechas L.), and fennel (Foeniculum vulgare) on common foodborne pathogens. *Journal of Agricultural and Food Chemistry*, 52, 8255–8260.

Dalziel, J. M. (1936). *The Useful Plants of West Tropical Africa*. London: The Crown Agents for the Overseas Governments and Administrations.

Davis, P. H. (1967). *Flora of Turkey and the Aegean Islands* (Vol. 2). Edinburgh: University Press.

Denev, R. V., Kuzmanova, I. S., Momchilova, S. M., & Nikolova-Damyanova, B. M. (2011). Resolution and quantification of isomeric fatty acids by silver ion HPLC: fatty acid composition of aniseed oil (Pimpinella anisum, Apiaceae). *Journal of AOAC International*, 94(1), 4–8.

Di Carlo G., Autore, G., Izzo, A. A., Maiolino, P., Mascolo, N., Viola, P., Diurno, M. V., & Capasso, F. (1993). Inhibition of intestinal motility and secretion by flavonoids in mice and rats: structure activity relationships. *Journal of Pharmacy and Pharmacology*, 45, 1054–1059.

Diaaz-Maroto, M. C., Hidalgo, I. J. D., Saanchez-Palomo, E., & Pearez Coello, M. S. (2005). Volatile components and key odorants of fennel (Foeniculum vulgare Mill.) and thyme (Thymus vulgaris L.) Oil extracts obtained by simultaneous distillation–extraction and supercritical fluid extraction. *Journal of Agricultural and Food Chemistry*, 53, 5385–5389.

Diaaz-Maroto, M. C., Pearez-Coello, M. S., Esteban, J., & Sanz, J. (2006). Comparison of the volatile composition of wild fennel samples (Foeniculum vulgare Mill.) from Central Spain. *Journal of Agricultural and Food Chemistry*, 54, 6814–6818.

Dığrak, M., Alma, M. H., & Ilçim, A. (2001). Antibacterial and antifungal activities of Turkish medicinal plants. *Pharmaceutical Biology*, 39, 346–350.

Dirks, U., & Herrmann, K. (1984). 4-(β-d-glucopyranosyloxy) benzoic acid, a characteristic phenolic constituent of the Apiaceae. *Phytochemistry*, 23(8), 1811–1812.

Dogan, M., & Akgul, A. (2005). Characteristics and fatty acid compositions of Rhus coriaria cultivators from south Turkey. *Chemistry of Natural Compounds*, 41, 724–725.

Drinkwater, N. R., Miller, E. C., Miller, J. A., & Pitot, H. C. (1976). Hepatocarcinogenicity of estragole (1-allyl-4-methoxybenzene) and 1-hydroxyestragole in the mouse and mutagenicity of 1 acetoxyestragole in bacteria. *Journal of the National Cancer Institute*, 57, 1323–1331.

Dulger, B., & Gonuz, A. (2004). Antimicrobial activity of some Turkish medicinal plants. *Pakistan Journal of Biological Sciences*, 7, 1559–1562.

El Ouariachi, E., Lahhit, N., Bouyanzer, A., Hammouti, B., Paolini, J., Majidi, L., … & Costa, J. (2014). Chemical composition and antioxidant activity of essential oils and solvent extracts of Foeniculum vulgare Mill. from Morocco. *Journal of Chemical and Pharmaceutical Research*, 6(4), 743–748.

El Sissi, H. I., Ishak, M. S., & Abd El Wahid, M. S. (1972). Polyphenolic components of Rhus coriaria leaves. *Planta Medica*, 21, 67–71.

El Sissi, H. I., Ishak, M. S., Abd El Wahid, M. S., & El Ansari, M. A. (1971). The gallotannins of Rhus coriaria and Mangifera indica. *Planta Medica*, 19, 342–351.

El Sissi, H. I., Saleh, N. A. M., & Abd El Wahid, M. S. (1966). The tannins of Rhus coriaria and Mangifera indica. *Planta Medica*, 14, 222–231.

Embong, M. B., Hadziyev, D., & Molnar, S. (1977). Essential oils from spices grown in Alberta. Anise oil (Pimpinella anisum). *Canadian Journal of Plant Science*, 57(3), 681–688.

Erler, F., Ulug, I., & Yalcinkaya, B. (2006). Repellent activity of five essential oils against Culex pipiens. *Fitoterapia*, 77(7–8), 491–494.

Esmaeili, A., & Asgari, A. (2015). In vitro release and biological activities of Carum copticum essential oil (CEO) loaded chitosan nanoparticles. *International Journal of Biological Macromolecules*, 81, 283–290.

European Commission, Opinion of the Scientific Committee on Food on Estragole (1-Allyl-4 methoxybenzene) (2001). European Commission Health & Consumer Protection Directorate-General SCF/CS/FLAV/FLAVOUR/6 ADD2 FI-NAL, Brussels, Belgium.

Faudale, M., Viladomat, F., Bastida, J., Poli, F., & Codina, C. (2008). Antioxidant activity and phenolic composition of wild, edible, and medicinal fennel from different mediterranean countries. *Journal of Agricultural and Food Chemistry*, 56, 1912–1920.

Fazeli, M. R., Amin, G., Attari, M. M. A., Ashtiani, H., Jamalifar, H., & Samadi, N. (2007). Antimicrobial activities of Iranian sumac and avishan-e shirazi (Zataria multiXora) against some food-borne bacteria. *Food Control*, 18, 646–649.

Fujimatu, E., Ishikawa, T., & Kitajima, J. (2003). Aromatic compound glucosides, alkyl glucoside and glucide from the fruit of anise. *Phytochemistry*, 63(5), 609–616.

Garg, B. D., Cadle, S. H., Mulawa, P. A., Groblicki, P. J., Laroo, C., & Parr, G. A. (2000). Brake wear particulate matter emissions. *Environmental Science & Technology*, 34(21), 4463–4469.

Giancarlo, S., Rosa, L. M., Nadjafi, F., & Francesco (2006). Hypoglycaemic activity of two spices extracts: Rhus coriaria L. and Bunium persicum Boiss. *Natural Product Research*, 20, 882–886.

Gilani, A. H., Bashir, S., & Khan, A. (2007). Pharmacological basis for the use of Borago officinalis in gastrointestinal, respiratory and cardiovascular disorders. *Journal of Ethnopharmacology*, 114, 393–399.

Gilligan, N. P. (2005). The palliation of nausea in hospice and palliative care patients with essential oils of Pimpinella anisum (aniseed), Foeniculum vulgare var. dulce (sweet fennel), Anthemis nobilis (Roman chamomile) and Mentha x piperita (peppermint). *International Journal of Aromatherapy*, 15(4), 163–167.

Gniewosz, M., Kraśniewska, K., Woreta, M., & Kosakowska, O. (2013). Antimicrobial activity of a pullulan–caraway essential oil coating on reduction of food microorganisms and quality in fresh baby carrot. *Journal of Food Science*, 78(8), M1242–M1248.

Grieve, M. (1931). *A Modern Herbal: The Medicinal, Culinary, Cosmetic and Economic Properties, Cultivation and Folk-lore of Herbs, Grasses, Fungi, Shrubs & Trees with Their Modern Scientific Uses*. San Diego, CA: Harcourt Brace & Company.

Gross, M., Lewinsohn, E., Tadmor, Y., Bar, E., Dudai, N., Cohen, Y., & Friedma, J. (2009). The inheritance of volatile phenylpropenes in bitter fennel (Foeniculum vulgare Mill. var. vulgare Apiaceae) chemotypes and their distribution within the plant. *Biochemical Systematics and Ecology*, 37, 308–316.

Gulcın, İ., Oktay, M., Kıreçcı, E., & Kufrevıoglu, Ö. İ. (2003). Screening of antioxidant and antimicrobial activities of anise (Pimpinella anisum L.) seed extracts. *Food Chemistry*, 83(3), 371–382.

Gündüz, G. T., Gönül, S. A., & Karapinar, M. (2010). Efficacy of sumac and oregano in the inactivation of Salmonella Typhimurium on tomatoes. *International Journal of Food Microbiology*, 141, 39–44.

Güvenç, A., & Koyuncu, M. (1994). A study on the main active compounds of leaves and fruits of Rhus coriaria L. *Turkish Journal of Medical Sciences*, 20, 11–13.

Hajlaoui, H., Arraouadi, S., Noumi, E., Aouadi, K., Adnan, M., Khan, M. A., ... & Snoussi, M. (2021). Antimicrobial, antioxidant, anti-acetylcholinesterase, antidiabetic, and pharmacokinetic properties of Carum carvi L. and Coriandrum sativum L. essential oils alone and in combination. *Molecules*, 26(12), 3625.

Hashem, M., & Alamri, S. (2010). Contamination of common spices in Saudi Arabia markets with potential mycotoxin-producing fungi. *Saudi Journal of Biological Sciences*, 17, 167–175.

Heidari, M. R., & Ayeli, M. (2005). Effects of methyl alcoholic extract of Pimpinella anisum L. on picrotoxin induced seizure in mice and its probable mechanism. *Scientific Journal of Kurdestan University of Medical Sicences*, 10, 1–8.

Iauk, L., Caccamo, F., Speciale, A. M., Tempera, G., Ragusa, S., & Pante, G. (1998). Antimicrobial activity of Rhus coriaria L. leaf extract. *Phytotherapy Research*, 12, 152–153.

Janahmadi, M., Farajnia, S., Vatanparast, J., Abbasipour, H., & Kamalinejad, M. (2008). The fruit essential oil of Pimpinella anisum L. (Umblliferae) induces neuronal hyperexcitability in snail partly through attenuation of after-hyperpolarization. *Journal of Ethnopharmacology*, 120(3), 360–365.

Javidnia, K., Dastgheib, L., Samani, S. M., & Nasiri, A. (2003). Antih irsutism activity of Fennel (fruits of Foeniculum vulgare) extract: a double-blind placebo controlled study. *Phytomedicine*, 10, 455–458.

Karami, N. K., Moattar, F., & Ghahiri, A. (2008). Comparison of effectiveness of an herbal drug (celery, saffron, Anise) and mephnamic acid capsule on primary dismenorrhea. *Ofoghe Danesh*, 14(1), 11–19.

Katzung, B. G., & Chatterjee, K. (2003). Vasodilators & the treatment of Angina Pectoris. In: Katzung, B. G., S. B. Masters, and A. J. Trevor. Eds., *Basic and Clinical Pharmacology* (9th Ed., pp. 187–192). New York, USA: McGraw-Hill Companies.

Kaur, G. J., & Arora, D. S. (2008). In-vitro antibacterial activity of three plants belonging to the family Umbelliferae. *International Journal of Antimicrobial Agents*, 31, 393–395.

Kaya, I., & Incekara, N. (2000). Contents of some wild plants species consumed as food in Aegean region. *Turkish Journal of Weed Science*, 3, 56–64.

Kazemipoor, M., Radzi, C. W. J. B. W. M., Hajifaraji, M., & Cordell, G. A. (2014). Preliminary safety evaluation and biochemical efficacy of a Carum carvi extract: results from a randomized, triple-blind, and placebo-controlled clinical trial. *Phytotherapy Research*, 28(10), 1456–1460.

Keshavarz, A., Minaiyan, M., Ghannadi, A., & Mahzouni, P. (2013). Effects of Carum carvi L. (Caraway) extract and essential oil on TNBS-induced colitis in rats. *Research in Pharmaceutical Sciences*, 8(1), 1.

Kim, D. H., Kim, S. I., Chang, K. S., & Ahn, Y. J. (2002). Repellent activity of constituents identified in Foeniculum vulgare fruit against Aedes aegypti (Diptera: Culicidae). *Journal of Agricultural and Food Chemistry*, 50, 6993–6996.

Kizil, S., & Turk, M. (2010). Microelement contents and fatty acid compositions of Rhus coriaria L. and Pistacia terebinthus L. fruits spread commonly in the south eastern Anatolia region of Turkey. *Natural Product Research*, 24, 92–98.

Kosalec, I., Pepeljnjak, S., & Kuštrak, D. (2005). Antifungal activity of fluid extract and essential oil from anise fruits (Pimpinella anisum L., Apiaceae). *Acta Pharmaceutica*, 55(4), 377–385.

Kosar, M., Bozan, B., Temelli, F., & Baser, K. H. C. (2007). Antioxidant activity and phenolic composition of sumac (Rhus coriaria L.) extracts. *Food Chemistry*, 103(3), 952–959.

Kreydiyyeh, S. I., Usta, J., Knio, K., Markossian, S., & Dagher, S. (2003). Aniseed oil increases glucose absorption and reduces urine output in the rat. *Life Sciences*, 74(5), 663–673.

Kumbasar, E. P. A., Atav, R., & Bahtiyari, M. I. (2009). Effects of alkali proteases on dyeing properties of various proteinous materials with natural dyes. *Textile Research Journal*, 79, 517–525.

Kunzemann, J., & Herrmann, K. (1977). Isolation and identification of flavon (ol)-O-glycosides in caraway (Carum carvi L.), fennel (Foeniculum vulgare Mill.), anise (Pimpinella anisum L.), and coriander (Coriandrum sativum L.), and of flavon-C-glycosides in anise: I. Phenolics of spices. *Zeitschrift für Lebensmittel-Untersuchung und Forschung*, 164, 194–200.

Kwon, Y. S., Choi, W. G., Kim, W. J., Kim, W. K., Kim, M. J., Kang, W. H., & Kim, C. M. (2002). Antimicrobial constituents of Foeniculum vulgare. *Archives of Pharmacal Research*, 25, 154–157.

Lauche, R., Janzen, A., Lüdtke, R., Cramer, H., Dobos, G., & Langhorst, J. (2015). Efficacy of caraway oil poultices in treating irritable bowel syndrome-a randomized controlled cross-over trial. *Digestion*, 92(1), 22–31.

Lee, H. S. (2004). p-Anisaldehyde: acaricidal component of Pimpinella anisum seed oil against the house dust mites Dermatophagoides farinae and Dermatophagoides pteronyssinus. *Planta Medica*, 70(03), 279–281.

Lee, J. B., Yamagishi, C., Hayashi, K., & Hayashi, T. (2011). Antiviral and immunostimulating effects of lignin-carbohydrate-protein complexes from Pimpinella anisum. *Bioscience, Biotechnology, and Biochemistry*, 75(3), 459–465.

Lee, S. H. (2004). Acaricidal activity of constituents identified in Foeniculum vulgare fruit oil against Dermatophagoides spp. (Acari: Pyroglyphidae). *Journal of Agricultural and Food Chemistry*, 52, 2887–2889.

Lev, E. (2002). Reconstructed materia medica of the Medieval and Ottoman al-Sham. *Journal of Ethnopharmacology*, 80, 167–179.

Lev, E. (2007). Drugs held and sold by pharmacists of the Jewish community of medieval (11, 14th centuries) Cairo according to lists of materia medica found at the Taylor–Schechter Genizah collection, Cambridge. *Journal of Ethnopharmacology*, 110, 275–293.

Lev, E., & Amar, Z. (2002). Ethnopharmacological survey of traditional drugs sold in the Kingdom of Jordan. *Journal of Ethnopharmacology*, 82, 131–145.

Lev, E., & Amar, Z. (2008). "Fossils" of practical medical knowledge from medieval Cairo. *Journal of Ethnopharmacology*, 119, 24–40.

Macrae, R., Robinson, R. K., & Sadler, M. J. (1993a). *Encyclopaedia of Food Science, Food Technology and Nutrition* (Vol. 5). San Diego, CA: Academic Press INC.

Macrae, R., Robinson, R. K., & Sadler, M. J. (1993b). *Encyclopaedia of Food Science, Food Technology and Nutrition* (Vol. 7). San Diego, CA: Academic Press INC.

Mahady, G. B., Pendland, S. L., Stoia, A., Hamill, F. A., Fabricant, D., Dietz, B. M., & Chadwick, L. R. (2005). In-vitro susceptibility of Helicobacter pylori to botanical extracts used traditionally for the treatment of gastro-intestinal disorders. *Phytotherapy Research*, 19, 988–999.

Mahdavi, S., Hesami, B., & Sharafi, Y. (2018). Antimicrobial and antioxidant activities of Iranian sumac (Rhus coriaria L.) fruit ethanolic extract. *Journal of Applied Microbiology and Biochemistry*, 2(5), 2576–1412.

Mardani, M., Afra, S. M., Tanideh, N., Andisheh Tadbir, A., Modarresi, F., Koohi-Hosseinabadi, O., … & Sepehrimanesh, M. (2016). Hydroalcoholic extract of Carum carvi L. in oral mucositis: a clinical trial in male golden hamsters. *Oral Diseases*, 22(1), 39–45.

Marino, S. D., Gala, F., Borbone, N., Zollo, F., Vitalini, S., Visioli, F., & Iorizzi, M. (2007). Phenolic glycosides from Foeniculum vulgare fruit and evaluation of antioxidative activity. *Phytochemistry*, 68, 1805–1812.

Maritim, A. C., Sanders, A., & Watkins Iii, J. B. (2003). Diabetes, oxidative stress, and antioxidants: a review. *Journal of Biochemical and Molecular Toxicology*, 17(1), 24–38.

Mavlyanov, S., Islambekov, S. Y., Ismailov, A., Karimdzhanou, A., & Ismalov, A. (1997). Anthocyans and organic acids of the fruits of some species of sumac. *Chemistry of Natural Compounds*, 33, 209.

McDougall, I., Brown, F. H., & Fleagle, J. G. (2005). Stratigraphic placement and age of modern humans from Kibish, Ethiopia. *Nature*, 433(7027), 733–736.

McLaughlin, M. M. J., Parker, D. R., & Clarke, J. M. (1999). Metals and micronutrients-food safety issues. *Field Crops Research*, 60, 143–163.

Mehrdad, M., Zebardast, M., Abedi, G., Koupaei, M. N., Rasouli, H., & Talebi, M. (2009). Validated high throughput HPLC method for the analysis of flavonol aglycones myricetin, quercetin, and kaempferol in Rhus coriaria L. using a monolithic column. *Journal of AOAC International*, 92, 1035–1043.

Miguel, M. G., Cruz, C., Faleiro, L., Simões, M. T., Figueiredo, A. C., Barroso, J. G., & Pedro, L. G. (2010). Foeniculum vulgare essential oils: chemical composition, antioxidant and antimicrobial activities. *Natural Product Communications*, 5(2), 319–328.

Miller, E. C., Swanson, A. B., Phillips, D. H., Fletcher, T. L., Liem, A., & Miller, J. A. (1983). Structure–activity studies of the carcinogenicities in the mouse and rat of some naturally occurring and synthetic alkenyl-benzene derivatives related to safrole and estragole. *Cancer Research*, 43, 1124–1134.

Miller, J. A., & Miller, E. C. (1983). The metabolic activation and nucleic acid adducts of naturally-occurring carcinogens: recent results with ethyl carbamate and the spice flavors safrole and estragole. *Brazilian Journal of Cancer*, 48, 1–15.

Miraj, S., & Kiani, S. (2016). Pharmacological activities of Carum carvi L. *Der Pharmacia Lettre*, 8(6), 135–138.

Mirheydar, H. (2001). *Herbal Information: Usage of Plants in Prevention and Treatment of Diseases* (pp. 12–19). Tehran, Iran: Islamic Culture Press Center.

MohammadAl-Ismail, K., & Aburjai, T. (2004). Antioxidant activity of water and alcohol extracts of chamomile flowers, anise seeds and dill seeds. *Journal of the Science of Food and Agriculture*, 84(2), 173–178.

Mohammadi, S., Kouhsari, S. M., & Feshani, A. M. (2010). Antidiabetic properties of the ethanolic extract of Rhus coriaria fruits in rats. *Daru*, 18(4), 270–274.

Mohsenzadeh, M. (2007). Evaluation of antibacterial activity of selected Iranian essential oils against Staphylococcus aureus and Escherichia coli in nutrient broth medium. *Pakistan Journal of Biological Sciences*, 10, 3693–3697.

Naghibi, S. M., Ramezani, M., Ayati, N., & Zakavi, S. R. (2015). Carum induced hypothyroidism: an interesting observation and an experiment. *DARU Journal of Pharmaceutical Sciences*, 23, 1–4.

Nahidi, F., Taherpoor, M., Mojab, F., & Majd, H. (2008). Effect of Anise extract on hot flush of menopause. *Pajoohandeh Journal*, 13(3), 167–73.

Nasar-Abbas, S. M., Halkman, A. K., & Al-Haq, M. I. (2004). Inhibition of some foodborne bacteria by alcohol extract of sumac (Rhus coriaria L.). *Journal of Food Safety*, 24, 257–267.

Oktay, M., Gulçin, İ., & Kufrevioğlu, Ö. İ. (2003). Determination of in vitro antioxidant activity of fennel (Foeniculum vulgare) seed extracts. *LWT-Food Science and Technology*, 36(2), 263–271.

Orav, A., Raal, A., & Arak, E. (2008). Essential oil composition of Pimpinella anisum L. fruits from various European countries. *Natural Product Research, 22*(3), 227–232.

Ostad, S. N., Soodi, M., Shariffzadeh, M., Khorshidi, N., & Marzban, H. (2001). The effect of fennel essential oil on uterine contraction as a model for dysmenorrhea, pharmacology and toxicology study. *Journal of Ethnopharmacology, 76*, 299–304.

Ozbek, H., Ugras, S., Dulger, H., Bayram, I., Tuncer, I., Ozturk, G., & Ozturk, A. (2003). Hepatoprotective effect of Foeniculum vulgare essential oil. *Fitoterapia, 74*, 317–319.

Özcan, M. (2003). Effect of sumach (Rhus coriaria L.) extracts on the oxidative stability of peanut oil. *Journal of Medicinal Food, 6*, 63–66.

Özcan, M. (2004). Mineral contents of some plants used as condiments in Turkey. *Food Chemistry, 84*, 437–440.

Özcan, M. M., & Akbulut, M. (2007). Estimation of minerals, nitrate and nitrite contents of medicinal and aromatic plants used as spices, condiments and herbal tea. *Food Chemistry, 106*, 852–858.

Ozcan, M. M., & Chalchat, J. C. (2006). Chemical composition and antifungal effect of anise (Pimpinella anisum L.) fruit oil at ripening stage. *Annals of Microbiology, 56*(4), 353–358.

Özcan, M., & Haciseferogullari, H. (2004). A condiment [Sumac (Rhus Coriaria L.) fruits]: some physio-chemical properties. *Journal of Plant Physiology, 30*, 74–84.

Pai, M. B., Prashant, G. M., Murlikrishna, K. S., Shivakumar, K. M., & Chandu, G. N. (2010). Antifungal efficacy of Punica granatum, Acacia nilotica, Cuminum cyminum and Foeniculum vulgare on Candida albicans: anin vitro study. *Indian Journal of Dental Research, 21*(3), 334–336.

Paini, A., Punt, A., Viton, F., Scholz, G., Delatour, T., Marin-Kuan, M., ... & Rietjens, I. M. (2010). A physiologically based biodynamic (PBBD) model for estragole DNA binding in rat liver based on in vitro kinetic data and estragole DNA adduct formation in primary hepatocytes. *Toxicology and Applied Pharmacology, 245*, 57–66.

Palombo, E. A. (2006). Phytochemicals from traditional medicinal plants used in the treatment of diarrhea: modes of action and effects on intestinal function. *Phytotherapy Research, 20*, 717–724.

Panico, A., Cardile, V., Santagati, N. A., & Messina, R. (2009). Antioxidant and protective effects of sumac leaves on chondrocytes. *Journal of Medicinal Plants Research, 3*, 855–861.

Parejo, I., Jauregui, O., Saa nchez-Rabaneda, F., Viladomat, F., Bastida, J., & Codina, C. (2004b). Separation and characterization of phenolic compounds in fennel (Foeniculum vulgare) using liquid chromatography–negative electrospray ionization tandem mass spectrometry. *Journal of Agricultural and Food Chemistry, 52*, 3679–3687.

Parejo, I., Viladomat, F., Bastida, J., Schmeda-Hirschmann, G., Burillo, J., & Codina, C. (2004a). Bioguided isolation and identifica tion of the nonvolatile antioxidant compounds from Fennel (Foeniculum vulgare Mill.) waste. *Journal of Agricultural and Food Chemistry, 52*, 1890–1897.

Park, H. J. (1996). Syringin 4-O-b-glucoside, a new phenylpropanoid glycoside, and costunolide, a nitric oxide synthase inhibitor, from the stem bark of Magnolia sieboldii. *Journal of Natural Products, 59*, 1128–1130.

Park, I. K., Choi, K. S., Kim, D. H., Choi, I. H., Kim, L. S., Bak, W. C., ... & Shin, S. C. (2006). Fumigant activity of plant essential oils and components from horseradish (Armoracia rusticana), anise (Pimpinella anisum) and garlic (Allium sativum) oils against Lycoriella ingenua (Diptera: Sciaridae). *Pest Management Science: Formerly Pesticide Science*, 62(8), 723–728.

Phillips, D. H. (1994). DNA adducts derived from safrole, estragole and related compounds, and from benzene and its metabolites. *IARC Scientific Publications*, 125, 131–140.

Picon, P. D., Picon, R. V., Costa, A. F., Sander, G. B., Amaral, K. M., Aboy, A. L., & Henriques, A. T. (2010). Randomized clinical trial of a phytotherapic compound containing Pimpinella anisum, Foeniculum vulgare, Sambucus nigra, and Cassia augustifolia for chronic constipation. *BMC Complementary and Alternative Medicine*, 10, 1–9.

Pingret, D., Fabiano-Tixier, A. S., & Chemat, F. (2013). Ultrasound-assisted extraction. In: M. A. Rostagno and J. M. Prado (eds.), *Natural Product Extraction: Principles and Applications*. The Royal Society of Chemistry, 89–112.

Pourahmad, J., Eskandari, M. R., Shakibaei, R., & Kamalinejad, M. (2010). A search for hepatoprotective activity of aqueous extract of Rhus coriaria L. against oxidative stress cytotoxicity. *Food and Chemical Toxicology*, 48, 854–858.

Pourgholami, M. H., Majzoob, S., Javadi, M., Kamalinejad, M., Fanaee, G. H. R., & Sayyah, M. (1999). The fruit essential oil of Pimpinella anisum exerts anticonvulsant effects in mice. *Journal of Ethnopharmacology*, 66(2), 211–215.

Pourmorad, F., Hosseinimehr, S. J., & Shahabimajd, N. (2006). Antioxidant activity, phenol and flavonoid contents of some selected Iranian medicinal plants. *African Journal of Biotechnology*, 5(11), 1142–1145.

Pradhan, M., Sribhuwaneswari, S., Karthikeyan, D., Minz, S., Sure, P., Chandu, A. N., ... & Sivakumar, T. (2008). In-vitro cytoprotection activity of Foeniculum vulgare and Helicteres isora in cultured human blood lymphocytes and antitumour activity against B16F10 melanoma cell line. *Research Journal of Pharmacy and Technology*, 1(14), 450–452.

Prajapati, V., Tripathi, A. K., Aggarwal, K. K., & Khanuja, S. P. S. (2005). Insecticidal, repellent and oviposition-deterrent activity of selected essential oils against Anopheles stephensi, Aedes aegypti and Culex quinquefasciatus. *Bioresource Technology*, 96(16), 1749–1757.

Punt, A., Freidig, A. P., Delatour, T., Scholz, G., Boersma, M. G., Schilter, B., ... & Rietjens, I. M. (2008). A physiolog ically based biokinetic (PBBK) model for estragole bioactivation and detoxification in rat. *Toxicology and Applied Pharmacology*, 231, 248–259.

Puri, M., Sharma, D., & Barrow, C. J. (2012). Enzyme-assisted extraction of bioactives from plants. *Trends in Biotechnology*, 30(1), 37–44.

Rajeshwari, U., Shobha, I., & Andallu, B. (2011). Comparison of aniseeds and coriander seeds for antidiabetic, hypolipidemic and antioxidant activities. *Spatula DD*, 1(1), 9–16.

Rayne, S., & Mazza, G. (2007). Biological activities of extracts from sumac (Rhus spp.): a review. *Plant Foods for Human Nutrition*, 62, 165–175.

Revuelta, M. P., Cantabrana, B., & Hidalgo, A. (1997). Depolarization dependent effect of flavonoids in rat uterine smooth muscle contraction elicited by $CaCl_2$. *General Pharmacology*, 29, 847–857.

Rodrigues, V. M., Rosa, P. T., Marques, M. O., Petenate, A. J., & Meireles, M. A. A. (2003). Supercritical extraction of essential oil from aniseed (Pimpinella anisum L) using CO_2: solubility, kinetics, and composition data. *Journal of Agricultural and Food Chemistry*, 51(6), 1518–1523.

Ruberto, G., Baratta, M. T., Deans, S. G., & Dorman, H. J. D. (2000). Antioxidant and antimicrobial activity of Foeniculum vulgare and Crithmum maritimum essential oils. *Planta Medica*, 66, 687–693.

Sabu, A., Sarita, S., Pandey, A., Bogar, B., Szakacs, G., & Soccol, C. R. (2002). Solid-state fermentation for production of phytase by Rhizopus oligosporus. *Applied Biochemistry and Biotechnology*, 102, 251–260.

Sadjadi, N. S., Shahi, M. M., Jalali, M. T., & Haidari, F. (2014). Short-term caraway extract administration improves cardiovascular disease risk markers in streptozotocin-induced diabetic rats: a dose-response study. *Journal of Dietary Supplements*, 11(1), 30–39.

Saeed, H. M. (1972). *Hamdard Pharmacographia Indica* (pp. 372–373). Pakistan: Hamdard National Foundation, Karachi.

Sağdiç, O., Karahan, A. G., Özcan, M., & Özkan, G. (2003). Effect of some spice extracts on bacterial inhibition. *Food Science and Technology International*, 9, 353–356.

Saghir, M. R., Sadiq, S., Nayak, S., & Tahir, M. U. (2012). Hypolipidemic effect of aqueous extract of Carum carvi (black Zeera) seeds in diet induced hyperlipidemic rats. *Pakistan Journal of Pharmaceutical Sciences*, 25(2), 333–337.

Sahraei, H., Ghoshooni, H., Salimi, S. H., Astani, A. M., Shafaghi, B., Falahi, M., & Kamalnegad, M. (2002). The effects of fruit essential oil of the Pimpinella anisum on acquisition and expression of morphine induced conditioned place preference in mice. *Journal of Ethnopharmacology*, 80(1), 43–47.

Said, O., Khalil, K., Fulder, S., & Azaizeh, H. (2002). Ethnopharmacological survey of medicinal Herbs in Israel, the Golan Heights and the West Bank region. *Journal of Ethnopharmacology*, 83, 251–265.

SalehiSurmaghi, M. H. (2010). *Medicinal Plants and Phytotherapy* (Vol. 1). Tehran, Iran: Donyay Taghziah Press.

Sen, S., Tascioglu, C., & Tırak, K. (2009). Fixation, leachability, and decay resistance of wood treated with some commercial extracts and wood preservative salts. *International Biodeterioration & Biodegradation*, 63, 135–141.

Seth, S. (2009). Calcium channel blockers. In: Seth, S. D. and V. Seth. Eds., *Textbook of Pharmacology* (3rd Ed., pp. 32–41). New Delhi: Elsevier.

Sezik, E., Yeşilada, E., Honda, G., Takaishi, Y., Takeda, Y., & Tanaka, T. (2001). Traditional medicine in Turkey X. Folk medicine in Central Anatolia. *Journal of Ethnopharmacology*, 75, 95–115.

Shahat, A. A., Ibrahim, A. Y., Hendawy, S. F., Omer, E. A., Hammouda, F. M., Rahman, F. H. A., & Saleh, M. A. (2011). Chemical com position, antimicrobial and antioxidant activities of essential oils from organically cultivated fennel cultivars. *Molecules*, 16, 1366–1377.

Shukla, H. S., Dubey, P., & Chaturvedi, R. V. (1989). Antiviral properties of essential oils of Foeniculum vulgare and Pimpinella anisum L. *Agronomie*, 9(3), 277–279.

Shukla, H. S., & Tripathi, S. G. (1987). Antifungal substance in the essential oil of anise (Pimpinella anisum L.). *Agricultural and Biological Chemistry*, 51(7), 1991–1993.

Singh, G., Kapoor, I. P. S., de Heluani, C. S., & Catalan, C. A. N. (2008). Chemical composition and antioxidant potential of essential oil and oleoresins from anise seeds (Pimpinella anisum L.).

Singh, G., Maurya, S., De Lampasona, M. P., & Catalan, C. (2006). Chemical constituents, antifungal and antioxidative potential of Foeniculum vulgare volatile oil and its acetone extract. *Food Control*, 17(9), 745–752.

Sofi, P. A., Zeerak, N. A., & Singh, P. (2009). Kala zeera (Bunium persicum Bioss.): a Kashmirian high value crop. *Turkish Journal of Biology*, 33(3), 249–258.

Sökmen, A., Jones, B. M., & Erturk, M. (1999). The in vitro antibacterial activity of Turkish medicinal plants. *Journal of Ethnopharmacology*, 67, 79–86.

Soliman, F. M., Shehata, A. F., Khaleel, A. E., & Ezzat, S. M. (2002). An acylated kaempferol glycoside from flowers of Foeniculum vulgare and F. dulce. *Molecules*, 7, 245–251.

Soylu, S., Yigitbas, H., Soylu, E. M., & Kurt, S. (2007). Antifungal effects of essential oils from oregano and fennel on Sclerotinia sclerotio rum. *Journal of Applied Microbiology*, 103, 1021–1030.

Speisky, H., Rocco, C., Carrasco, C., Lissi, E. A., & López-Alarcón, C. (2006). Antioxidant screening of medicinal herbal teas. *Phytotherapy Research: An International Journal Devoted to Pharmacological and Toxicological Evaluation of Natural Product Derivatives*, 20(6), 462–467.

Subbaraj, G. K., Kulanthaivel, L., Rajendran, R., & Veerabathiran, R. (2013). Ethanolic extract of Carum carvi (EECC) prevents N-nitrosodiethylamine induced phenobarbital promoted hepatocarcinogenesis by modulating antioxidant enzymes. *The International Journal of Pharmacy and Pharmaceutical Sciences*, 5, 195–9.

Subehan, Zaidi, S. F. H., Kadota, S., & Tezuka, Y. (2007). Inhibition on human liver cytochrome P450 3A4 by constituents of fennel (Foeniculum vulgare): identification and characterization of a mechanism-based inactivator. *Journal of Agricultural and Food Chemstry*, 55, 10162 10167.

Swanson, A. B., Chambliss, D. D., Blomquist, J. C., Miller, E. C., & Miller, J. A. (1979). The mutagenicities of safrole, estragole, eugenol, trans anethole, and some of their known or possible metabolites for Salmonella typhimurium mutants. *Mutation Research*, 60, 143–153.

Swanson, A. B., Miller, E. C., & Miller, J. A. (1981). The side-chain epoxidation and hydroxylation of the hepatocarcinogens safrole and estragole and some related compounds by rat and mouse liver microsomes. *Biochimica et Biophysica Acta*, 673, 504–516.

Tas, A. (2009). Analgesic effect of Pimpinella anisum L. essential oil extract in mice. *The Indian Veterinary Journal*, 86(2), 145–147.

Tas, A., Özbek, H., Atasoy, N., Altug, M. E., & Ceylan, E. (2006). Evaluation of analgesic and anti-inflammatory activity of Pimpinella anisum fixed oil extract. *The Indian Veterinary Journal*, 83(8), 840–843.

Telci, I., Demirtas, I., & Sachin, A. (2009). Variation in plant properties and essential oil composition of sweet fennel (Foeniculum vulgare Mill.) fruit during stages of maturity. *Industrial Crops and Products*, 30, 126–130.

Tirapelli, C. R., de Andrade, C. R., Cassano, A. O., De Souza, F. A., Ambrosio, S. R., da Costa, F. B., & de Oliveira, A. M. (2007). Antispasmodic and relaxant effects of the hidroalcoholic extract of Pimpinella anisum (Apiaceae) on rat anococcygeus smooth muscle. *Journal of Ethnopharmacology*, 110(1), 23–29.

Tognolini, M., Ballabeni, V., Bertoni, S., Bruni, R., Impicciatore, M., & Barocelli, E. (2007). Protective effect of Foeniculum vulgare essential oil and anethole in an experimental model of thrombosis. *Pharmacological Research*, 56, 254–260.

Topal, U., Sasaki, M., Goto, M., & Otles, S. (2008). Chemical compositions and antioxidant properties of essential oils from nine species of Turkish plants obtained by supercritical carbon dioxide extraction and steam distillation. *International Journal of Food Sciences and* Nutrition, 59(7–8), 619–634.

Tunç I., Berger, B. M., Erler, F., & Dagli, F. (2000). Ovicidal activity of essential oils from five plants against two stored-product insects. *Journal of Stored Products Research*, 36(2), 161–168.

Tuzlacı, E., & Aymaz, P. E. (2001). Turkish folk medicinal plants, Part IV: Gönen (Balıkesir). *Fitoterap*, 72, 323–343.

Twaij, H. A., Elisha, E. E., Khalid, R. M., & Paul, N. J. (1987). Analgesic studies on some Iraqi medicinal plants. *International Journal of Crude Drug Research*, 25(4), 251–254.

Usmanghani, K., Saeed, A., & Alam, M. T. (1997). *Indusyunic Medicine* (pp. 378–379). Karachi: University of Karachi Press.

Van Loo, P., De Bruyn, A., & Verzele, M. (1988). On the liquid chromatography and identification of the flavonoids, present in the "Sumac Tannic Acid" extracted from Rhus coriaria. *Chromatograph*, 25, 15–20.

Veggi, P. C., Cavalcanti, R. N., & Meireles, M. A. A. (2011). Modifier effects on supercritical fluid extraction (SFE) of some Brazilian plants: antioxidant activity and economical evaluation. *Procedia Food Science*, 1, 1717–1724.

Wilson, D. S. (1998). Adaptive individual differences within single populations. *Philosophical Transactions of the Royal Society of London. Series B: Biological Sciences*, 353(1366), 199–205.

Yaniv, Z. (1982). Medicinal plants in the perspective of time. *Mada*, 26, 10–22 (in Hebrew).

Yazdani, D., Rezazadeh, S. H., Amin, G. H., Abidin, Z., Shahnazi, S., & Jamalifar, H. (2009). Antifungal activity of dried extracts of anise (Pimpinella anisum L.) and star anise (Illicium verum Hook. f.) against dermatophyte and saprophyte fungi. *Journal of Medicinal Plants*, 8(29), 24–29.

Yeasmen, N., & Islam, M. N. (2015). Ethanol as a solvent and hot extraction technique preserved the antioxidant properties of tamarind (Tamarindus indica) seed. *Journal of Advanced Veterinary and Animal Research*, 2(3), 332–337.

Zalacain, A., Prodanov, M., Carmona, M., & Alonso, G. L. (2003). Optimisation of extraction and identification of gallotannins from sumac leaves. *Biosystems Engineering*, 84, 211–216.

Zargari, R. (1996). *Medicinal Plants*. Tehran, Iran: Tehran University Press.

Zargham, H., Zargham, H., & Zargham, R. (2008). Tannin extracted from sumac inhibits vascular smooth muscle cell migration. *McGill Journal of Medicine*, 11, 119–123.

Zeller, A., & Rychlik, M. (2006). Character impact odorants of fennel fruits and fennel tea. *Journal of Agricultural and Food Chemistry*, 54, 3686–3692.

Zheng, X., Liu, B., Li, L., & Zhu, X. (2011). Microwave-assisted extraction and antioxidant activity of total phenolic compounds from pomegranate peel. *Journal of Medicinal Plants Research*, 5(6), 1004–1011.

chapter five

Therapeutic potential of medicinal fruits of Lebanon

Meenakshisundaram Selvamuthukumaran

5.1 Introduction

Lebanon is commercially cultivating a lot of medicinal fruits, including pomegranate, fig, kiwi, jujuba, etc. The medicinal properties of these fruits are described below.

5.2 Pomegranate

Pomegranate, scientifically known as *Punica granatum*, is grown widely in Lebanon and other Middle Eastern countries. It's a small tree that can grow up to a height of 10 m. The husk portion of the fruit is red-purple in color, possessing an outer hard pericarp and mesocarp, i.e., the white albedo portion, which is spongy in nature and constitutes the inner wall of the fruit, where the seeds are attached (Stover and Mercure, 2007). Juice can be obtained from the mature fruits by crushing the seeds, yielding juice with a tart flavor and lower pH. The juice pressed from the pomegranate fruit contains higher levels of polyphenols (Fernandes et al., 2015).

5.2.1 Bio-active constituents of pomegranate fruit

The pomegranate fruit is a rich source of bioactive components, ranging from a wide variety of polyphenols to anthocyanins, as provided in Table 5.1.

5.2.2 Therapeutic properties of pomegranate fruit

Pomegranate is a medicinal fruit that contains excellent therapeutic activities, which can be ascribed to the presence of a wide variety of polyphenolic compounds. All parts of the fruit have been found to possess good pharmacological properties, including anti-inflammatory, anticancer, antiviral, antioxidant, anti-diabetic, and anti-hepatotoxic properties

 DOI: 10.1201/9781003452614-5

Table 5.1 Various bio-active constituents of pomegranate fruit

Name of the component	Bio-active constituents present	References
Juice	Polyphenol viz punicalagins; anthocyanins viz. cyanidin, glycosides of pelargonidin and delphinidin	Gómez Caravaca et al. (2013) and Hernández et al. (1999)
Peel	Condensed tannins, catechins and prodelphinidins	Nasr (1996) and Plumb et al. (2002)
Seed oil	Palmitic acid, stearic acid, punicic acid, oleic and linoleic acid	Schubert et al. (1999)

(Kumar et al., 2010). They also help enhance skin and cardiovascular health, prevent Alzheimer's disease, and improve sperm quality (Bae et al., 2010; Turk et al., 2008; Cam et al., 2009; Aslam et al., 2006; Forest et al., 2007; Singh et al., 2008; Hamad and Al-Momene, 2009; Davidson et al., 2009; Duman et al., 2009; Lee et al., 2010).

5.3 *Fig*

Scientifically, it is known as *Ficus carica*, growing widely in Lebanon. It's a small tree that yields edible fruit and belongs to the family Moraceae. The fruit can be eaten as such, or it can be processed to obtain fruit jam, or it can be dehydrated in order to reduce the moisture content and increase the stability. The raw fruit contains a potassium content of around 230 mg and dietary fiber of 3 g/100g.

The fruit contains several phytoconstituents, which include polyphenols like catechin, rutin, epicatechin, gallic acid, syringic acid, and chlorogenic acid (Veberic et al., 2008). The presence of anthocyanins, viz. cyanidin-3-O-rutinoside, can create variations in the color of different cultivars (Solomon et al., 2006).

5.3.1 *Therapeutic properties of fig fruit*

In the olden days, in Mediterranean countries, the milky sap obtained from the fig plant has been traditionally used for removing warts, softening calluses, and deterring parasites (Landranco, 2001). In the 18th century, syrup made from fig, combined with senna, was used as a laxative. The therapeutic benefits of consuming fig fruits are given in Figure 5.1.

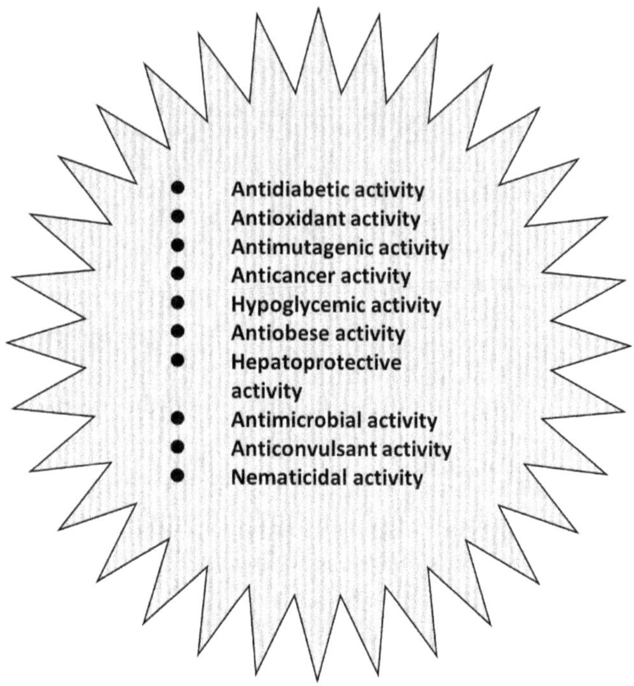

Figure 5.1 Therapeutic benefits of fig fruit consumption.

5.4 Jujuba

Its scientific name is *Ziziphus jujuba* and is commonly called red dates, growing widely in Lebanon. It belongs to the family Rhamnaceae. The plant is a deciduous tree that can grow up to a height of 10–12 m. The jujuba contains active phytoconstituents, and the leaves of the plant possess constituents like ziziphin and saponin, which can suppress the ability to perceive sweetness (Kurihara, 1992). The fruit obtained from this plant possesses terpenoids like alphitolic acid, colubrinic acid, and flavonoids like Quercetine 3-*O*-rutinoside, Kaempferol 3-*O*-rutinoside, and quercetin 3-*O*-robinobioside (Mahajan and Chopda, 2017). The fruit is eaten as such, or it can be consumed as a dried, candied fruit during coffee time. The nutritional constituents of jujube are given in Figure 5.2. The fruits and seeds obtained from jujuba have been used in traditional medicine to alleviate stress (Goetz, 2009) and also for sedation (Jian et al., 2007). It is used as an antifungal or antiseptic agent, mainly to control fungus. It has good anti-inflammatory activity, regulates blood pressure, enhances immunity, prevents ulcers, and helps in healing wounds. The fruits of jujuba are also combined with other herbs for efficiently curing colds and influenza. It is also used to heal and protect internal organs like the kidney, spleen, and heart.

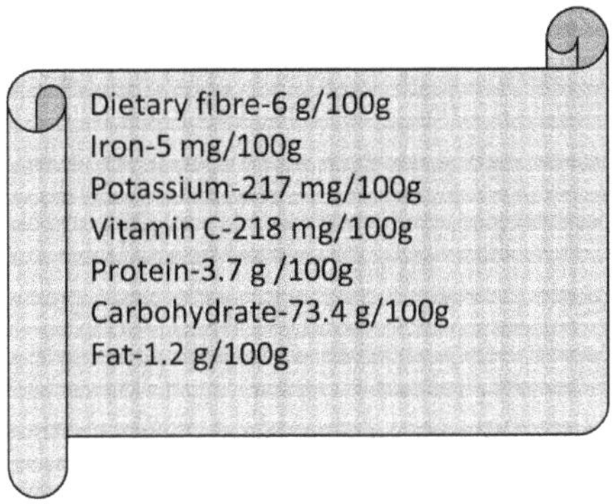

Figure 5.2 Nutritional constituents of Jujuba fruit on dry weight basis.

5.5 *Kiwi*

Scientifically, they are known as *Actinidia deliciosa*, a woody vine belonging to the family Actinidiaceae. The plant is widely cultivated in Lebanon, and the fruit possesses a slight acidic taste. It can be consumed raw or in processed form. The fruit contains 94 mg/100 g of vitamin C, 41 µg/100 g of vitamin K, 1.4 mg/100 g of vitamin E, 3 g/100 g of dietary fiber, 1 g/100 g of protein, and 14 g/100 g of carbohydrates. The major therapeutic benefits associated with consuming kiwi fruit include antioxidant activity, anti-inflammatory activity, enhanced skin health, prevention of hair fall, reduction of constipation, and improved digestion.

5.6 *Conclusions*

There are wide variety of medicinal fruits that can be commercially cultivated and utilized in Lebanon and the consumer can reap the benefit, if they consume such fruits as part of their daily diet.

References

Aslam MN, Lansky EP, Varani J. Pomegranate as a cosmeceutical source: Pomegranate fractions promote proliferation and procollagen synthesis and inhibit matrix metalloproteinase-1 production in human skin cells. *Journal of Ethnopharmacology*. 2006; 103(3):311–318.

Bae JY, Choi JS, Kang SW, Lee YJ, Park J, Kang YH. Dietary compound ellagic acid alleviates skin wrinkle and inflammation induced by UV-B irradiation. *Experimental Dermatology*. 2010; 19(8):82–90.

Cam M, Hısıl Y, Durmaz G. Classification of eight pomegranate juices based on antioxidant capacity measured by four methods. *Food Chemistry*. 2009; 112(3):721–726.

Davidson MH, Maki KC, Dicklin MR, Feinstein SB, Witchger MS, Bell M, McGuire DK, Provost J-C, Liker H, Aviram M. Effects of consumption of pomegranate juice on carotid intimamedia thickness in men and women at moderate risk for coronary heart disease. *International Journal of Oncology*. 2009; 104(7):936–942.

Duman AD, Ozgen M, Dayisoylu KS, Erbil N, Durgac C. Antimicrobial activity of six pomegranate (*Punica granatum* L.) varieties and their relation to some of their pomological and phytonutrient characteristics. *Molecules*. 2009; 14(5):1808–1817.

Fernandes L, Pereira JA, López Cortés I, Salazar DM, Ramalhosa EC. Physicochemical changes and antioxidant activity of juice, skin, pellicle and seed of pomegranate (cv. Mollar de Elche) at different stages of ripening. *Food Technology and Biotechnology*. 2015; 53(4):397–406.

Forest CP, Padma-Nathan H, Liker HR. Efficacy and safety of pomegranate juice on improvement of erectile dysfunction in male patients with mild to moderate erectile dysfunction: A randomized, placebo-controlled, double-blind crossover study. *International Journal of Impotence Research*. 2007; 19(6):564–567.

Goetz P. Mise en évidence d'un effet psychotrope de la teinture mère de Zizyphus jujuba Mill. *Phytothérapie*. 2009; 7:31–36.

Gómez Caravaca AM, Verardo V, Toselli M, Segura Carretero A, Fernández Gutiérrez A, and Caboni MF. Determination of the major phenolic compounds in pomegranate juices by HPLC-DAD-ESI-MS. *Journal of Agricultural and Food Chemistry*. 2013; 61(22):5328–5337.

Hamad AW, Al-Momene W. Separation and purification of crude ellagic acid from white flesh of pomegranate fruits as a potent anti-carcinogenic. *New Biotechnology*. 2009; 25(1):286.

Hernández F, Melgarejo P, Tomás-Barberán FA, Artés F. Evolution of juice anthocyanins during ripening of new selected pomegranate (Punica granatum) clones. *European Food Research and Technology*. 1999; 210(1):39–42.

Jiang J-G, Huang X-J, Chen J, Lin Q-S. Comparison of the sedative and hypnotic effects of flavonoids, saponins, and polysaccharides extracted from Semen Ziziphus jujube. *Natural Product Research*. 2007; 21(4):310–320.

Kumar N, Neeraj, Kumar S. Functional properties of pomegranate (Punica granatum L.). *The Pharma Innovation Journal*, 2010; 7(10):71–81.

Kurihara Y. Characteristics of antisweet substances, sweet proteins, and sweetness-inducing proteins. *Critical Reviews in Food Science and Nutrition*. 1992; 32(3):231–252.

Landranco G (2001). *Mediċina popolari ta' l-imgħoddi fil-gżejjer Maltin [Popular medicine of the past in the Maltese islands]* (in Maltese). Valletta, Malta: Klabb Kotba Maltin.

Lee CJ, Chen LG, Liang WL, Wanga CC. Anti-inflammatory effects of *Punica granatum* Linne *in vitro* and *in vivo*. *Food Chemistry*. 2010; 118(2):315–322.

Mahajan R, Chopda M. Phyto-pharmacology of Ziziphus jujuba Mill: A plant review. *Pharmacognosy Reviews*, 2017; 3(6):320–329.

Nasr CB. Quantitative determination of the polyphenolic content of pomegranate peel. *Zeitschrift für Lebensmittel-Untersuchung und Forschung*. 1996; 203(4):374–378.

Plumb GW, De Pascual-Teresa S, Santos-Buelga C, Rivas-Gonzalo JC, Williamson G. Antioxidant properties of gallocatechin and prodelphinidins from pomegranate peel. *Redox Report*. 2002; 7(41):41–46.

Schubert SY, Lansky EP, Neeman I. Antioxidant and eicosanoid enzyme inhibition properties of pomegranate seed oil and fermented juice flavonoids. *Journal of Ethnopharmacology*. 1999; 66(1):11–17.

Singh M, Arseneault M, Sanderson T, Morthy V, Ramassamy C. Challenges for research on polyphenols from foods in Alzheimer's disease: Bioavailability, metabolism and cellular and molecular mechanism. *Journal of Agricultural and Food Chemistry*. 2008; 56(13):4855–4873.

Solomon A, Golubowicz S, Yablowicz Z, Grossman S, Bergman M, Gottlieb HE, Altman A, Kerem Z, Flaishman MA. Antioxidant activities and anthocyanin content of fresh fruits of common fig (Ficus carica L.). *Journal of Agricultural and Food Chemistry*. 2006; 54(20):7717–7723.

Stover E, Mercure EW. The pomegranate: A new look at the fruit of paradise. *HortScience*. 2007; 42(5):1088–1092.

Turk G, Sonmez M, Aydin M, Yuce A, Gur S, Yuksel M, Aksu EH, Aksoy H. Effects of pomegranate juice consumption on sperm quality, sperm at genic cell density, antioxidant activity and testosterone level in male rats. *Clinical Nutrition*. 2008; 27(2):289–296.

Veberic R, Colaric M, Stampar F. Phenolic acids and flavonoids of fig fruit (Ficus carica L.) in the northern Mediterranean region. *Food Chemistry*. 2008; 106(1):153–157.

chapter six

Value addition of medicinal fruits of Muscat

Divya Chauhan and Sachin Kumar

6.1 Introduction

Since ancient times, medicinal fruits have been recognized for their therapeutic properties. They are regarded as potential sources of lead compounds for drug development as well as pharmacological agents (Lulekal et al., 2013). The World Health Organization states that over 80% of people living in poor nations receive their primary care from herbal and other traditional medicine practitioners. Secondary plant metabolites are responsible for the beneficial effects of plant materials (ürstenberg-Hägg et al., 2013). Oman is blessed with a vast array of therapeutic plants from nature (Ghazanfar and Al Sabahi, 1993). Therefore, screening and investigating these traditional medicinal herbs for bioactive compounds is crucial to lay the groundwork for future pharmacological research (Hemalatha et al., 2013). Therefore, screening and investigating these traditional medicinal herbs for bioactive compounds is crucial to lay the groundwork for future pharmacological research (Koehn, 2009). Value addition can be defined as an additional characteristic of an object of interest, which could be a person, service, product, or anything else that goes above and beyond what is typically expected, even at a higher cost to the customer. Value addition, in short, is the process of giving a food product (raw material) additional value in order to make it into a more valuable product. A food product can be made more valuable by the food industry in several ways, including processing, packaging, genetic manipulation, and the addition of new flavors and colors. The growing health and nutrition demands of consumers have made it possible for companies to create goods that meet those needs, thanks to technological advancements. Additionally, manufacturers create items intended for final consumers. This entails doing more than just creating raw materials (Coltrain, 2000) (Figure 6.1).

DOI: 10.1201/9781003452614-6

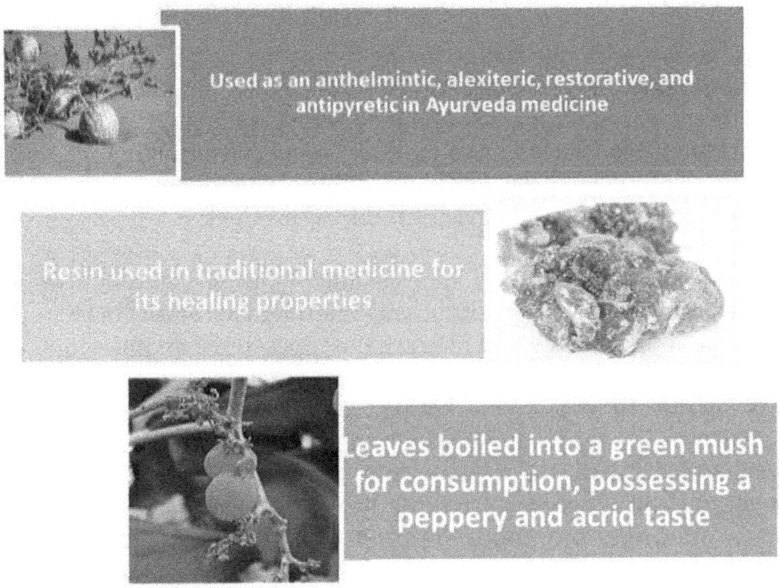

Figure 6.1 The fruit and vegetables global value chain.

6.1.1 *Importance of value addition*

Due to fierce competition in international marketplaces, farmers are being advised to incorporate value-adding tactics into their products to remain competitive, which has led to a decrease in farm revenue. Post-harvest operations and crop value addition, which boost farm income, strengthen the state economy, and achieve food security, are among the crucial components that support better agricultural production and strengthen ties between farmers and markets. According to Evans, value addition is an activity that farmers can use to create a new commodity by altering its location and timing, shifting it from one set of traits to another set of characteristics that are more desirable in the market in order to increase returns.

6.1.2 *Agricultural marketing resource centre (AMRC) described value addition as*

- Changing the physical form or state of an agricultural product, such as milling wheat to make flour, corn to make ethanol, and canola to make biodiesel.
- Making modifications to the production process and post-harvest handling that increase the product's value.
- Marketing based on distinctive features, such as selling soybeans through a system that remains intact.

6.1.3 The USDA defined value addition in agriculture as

i. When an agricultural commodity's physical state is altered, more money is made from marketing, processing, and physical segregation.
ii. These techniques include changes in physical condition, production in a way that increases the product's value, physical segregation, the source of a farm's base renewable energy, and marketing the product as a locally produced agricultural food product.

In agriculture, there are two approaches to adding value that improve the margin between gross output and intermediate input costs: improving production efficiency and altering the types, uses, qualities, and attributes of food (Lambert et al., 2006). Two distinct approaches are used to assess the effectiveness of marketing, namely innovation and coordination, when it comes to value addition in food items. In the first example, a novel and distinctive product is introduced to the market. Innovation is primarily concerned with making improvements to current practices, policies, offerings, and goods in order to produce new and improved products. Innovation can be achieved by researching and developing alternative crops (like castor beans for their oil) in place of conventional ones. Value-added producers could increase their earnings by developing novel goods and alternative crops. The selling of agricultural products and the arrangements for their production are the primary foci of coordination. It modifies the normal marketing exchanges between food manufacturers, suppliers, retailers, and consumers. Their cooperation results in higher-quality and more potent final goods. High-quality items cannot be produced by individual effort alone. Because different people have varied skills and talents to develop and market a product, joint effort is essential to increase market competency and reduce budget (Coltrain, 2000) (Figure 6.2).

6.1.4 Medicinal fruit crops of Oman

Dates (*Phoenix dactylifera*): Dates are a staple in Omani cuisine and have been used for centuries for their health benefits. They are rich in fiber, vitamins, and minerals and are known to support digestion, heart health, and energy levels.

Pomegranate (*Punica granatum*): Pomegranates are valued for their antioxidant properties. They are believed to help reduce inflammation, improve heart health, and may have anti-cancer properties.

Fig (*Ficus carica*): Figs are known for their high fiber content and various vitamins and minerals. They are used traditionally to aid digestion, promote bone health, and manage blood sugar levels.

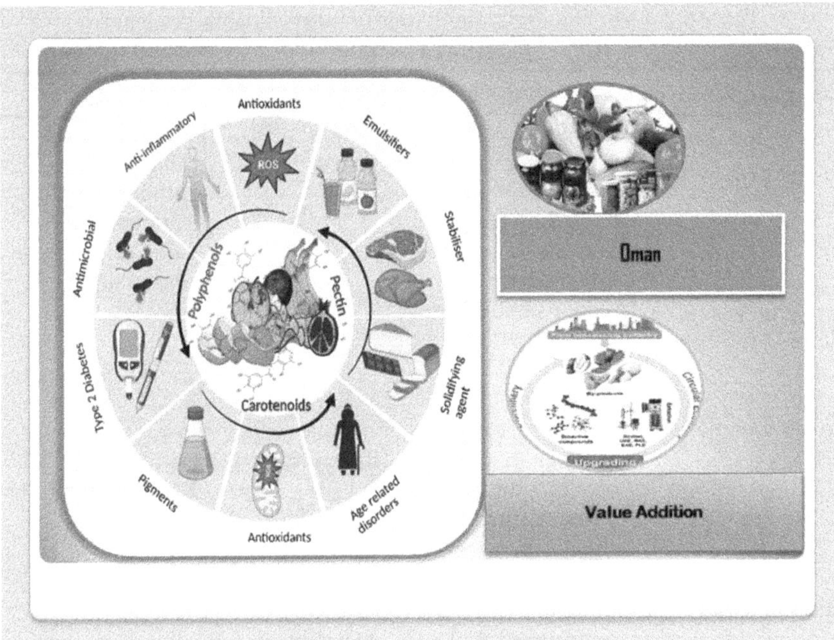

Figure 6.2 Value addition of fruits.

Jujube (*Ziziphus jujuba*): Jujube fruit is used in traditional medicine for its calming effects. It's believed to help with sleep disorders, digestive issues, and to have anti-anxiety properties (Table 6.1).

6.1.5 *Nutritional and biological evaluation of some traditionally important wild edible fruits of Muscat region*

6.1.5.1 *Dates (*Phoenix dactylifera*)*

6.1.5.1.1 Nutritional characteristics of dates fruits of Muscat The 6.5%–11.5% total dietary fiber (up to 90% insoluble and 10% soluble), around 1% fat, 2% proteins, and 2% ash have all been found to be present in date fruits. Additionally, phenolic antioxidants are abundant in them (Hong et al., 2006). Similarly, whereas the dry date fruit has a large amount of sucrose, the soft date fruit is primarily made up of invert sugars (fructose and glucose) with little to no sucrose. Thus, the fruits are categorized according to the type of sugar they contain: (i) invert sugar types, which are primarily composed of glucose and fructose (e.g., Barhi and Saidy); (ii) mixed sugar types, which are primarily composed of sucrose (e.g., Deglet Nour and Deglet Beidha); and (iii) cane sugar types (Ghnimi et al., 2017). Date fruits are highly healthy since they are packed with a variety of vital

Table 6.1 List of biological activities of nutraceuticals present in fruits with examples

Biological activity	Definition	Nutraceuticals	Examples
Anti-oxidant	Group of compounds that prevent damage caused by free radicals to the cells. They reduce oxidative stress and help in maintaining the cellular integrity.	Phenolic acids, flavonoids	Flavonols, catechins, anthocyanins, isoflavonoids, gallic acid, vanillic, ellagic, etc.
Anti-inflammatory	Inhibit or reduce inflammation by blocking the body's inflammatory signals or chemicals	Phenolic acids, flavonoids	Iso-vitexin, iso-orientin, 11-deoxycucurbita cinI-2-*O*-β-ᴅ
Anti-microbial	Group of compounds that kill or inhibit the growth of bacteria, fungi, and viruses	Alkaloids, cucurbitacin, terpenes, terpenoids, phenols	Choline, colocynthosides A, curcubitacin B, thymol, linalool, geraniol, linalyl acetate, carvacrol, and menthol
Anti-tumor	Chemicals or compounds that target and eliminate tumor cells or prevent their growth and spread	Cucurbitacin, saponin, flavonoids	Iso-orientin 3′-methylether, colocynthosides A, curcubitacin B
Anti-diarrheal	Compounds are used in reducing the severity of diarrhea	Phenolic compounds, flavonoids, glycosides, steroids	Isoflavonoids, cucurbitacin E 2-*O*-ᴅ-glucopyrano side.
Anti-allergic	Class of medication used to alleviate symptoms associated with allergic reactions	Glycosides	Cucurbitacin E 2-*O*-ᴅ-glucopyrano side, cucurbitacin E
Anti-mutagenic	Class of chemicals that protect cells from mutagenesis	Polyphenolic compounds	Quercetin, catechins

elements. According to Juices (2018), the dried fruits are mostly composed of sugar (80%), with lesser levels of protein, fiber, and trace elements such as boron, cobalt, copper, fluorine, magnesium, manganese, selenium, and zinc. There have been attempts to create functional foods from date fruits because of their high antioxidant and nutritional content (Selim et al., 2012).

6.1.6 *Antioxidant and anuimicrobial properties of date fruits of Muscat*

In many cultures, date fruit is used as traditional medicine to cure conditions like bronchitis, fever, intestinal issues, and wound healing. Their abundance of vital nutrients is the reason for this. More proof that date fruits can be used medicinally has come from recent investigations. Preclinical research has demonstrated the antibacterial and anti-inflammatory properties of date fruit. However, further research is still required to assess the therapeutic validity of these findings, particularly human trials. In a similar vein, research on the chemicals' mode of action in date fruits is necessary (Taleb et al., 2016).

6.1.7 *Pomegranate* (Punica granatum)

6.1.7.1 *Nutritional characteristics of dates fruits of Muscat*

There are civilizations that employ date fruit as traditional medicine to cure conditions like wound healing, fever, bronchitis, and intestinal issues. This is a result of the wide range of vital nutrients they contain. More proof of the medical benefits of date fruits has come from recent studies. Date fruit thus possesses antibacterial and anti-inflammatory properties, according to preclinical research. However, more research is still required to assess the therapeutic validity of these findings, particularly human trials. According to Taleb et al. (2016), further research is necessary to determine the chemicals' mode of action in date fruits.

6.1.8 *Antioxidant and antimicrobial properties of pomegranate fruits of Muscat*

Pomegranate (*Punica granatum*) has shown numerous health benefits, such as antioxidant, anti-inflammatory, and antimicrobial activities. The special phytochemical components of pomegranate peels are responsible for these health benefits. The phytochemical content and composition of pomegranate fruits and peels, as well as their potential health effects, are greatly influenced by variations in pomegranate cultivar, geographic location, and extraction techniques. Thus, the purpose of this study was to investigate the antioxidant and antibacterial properties of the phytochemicals found in Jordanian pomegranate peels. Due to its antibacterial and antioxidant properties, pomegranate peel extracts have recently been explored as possible natural food preservatives (El-Hamamsy and Khamissi, 2020).

6.1.9 *Fig* (Ficus carica)

6.1.9.1 *Nutritional characteristics of fig fruits of Muscat*

Among the phytochemicals present in figs, two main types are polyphenols and carotenoids. Phenolic acids, flavones, flavonones, flavonols, anthocyanins, and proanthocyanidins are the primary polyphenolic groups found in figs. Figs have a higher phenolic content than red wine and tea, which are two well-known and extensively researched sources of various phenolic compounds (Vinson, 1999).

6.1.10 *Antioxidant and antimicrobial properties of fig fruits of Muscat*

Common secondary metabolites found in plants, phenolic compounds have both physiological and beneficial impacts on human health due to their antioxidant properties (Çalişkan and Polat, 2011). Antioxidants help strengthen the immune system and are crucial in preventing pathogenic processes linked to asthma, cancer, heart disease, macular degeneration, and cataracts. According to Nakilcioğlu and Hışılu (2013), antioxidant defenses protect the body from the damaging effects of free radicals, which are produced as byproducts of normal metabolism.

6.1.11 *Jujube* (Ziziphus jujuba)

6.1.11.1 *Nutritional characteristics of Jujube* (Ziziphus jujuba) *fruits of Muscat*

Jujube fruit contains abundant nutrients, including polysaccharides, amino acids, ascorbic acid, triterpenic acids, flavonoids, phenolic acids, and mineral components. Furthermore, prior research has demonstrated that jujube fruit's numerous biological activities, including its antibacterial, anti-inflammatory, antioxidant, and anticancer qualities, may help treat a variety of illnesses. In the past, jujube fruit was consumed both fresh and dried. Fresh jujube is preferred because of its higher nutrient content and better flavor; however, because of its short shelf life, fresh jujube development has received less attention. Over the past three decades, there has been a surge in the production and availability of fresh jujube due to advancements in marketing logistics and jujube fruit preservation technologies.

Moreover, previous scientific evidence has revealed that jujube fruit has the potential to treat different diseases due to its various biological activities, such as antioxidant activity, anti-inflammatory, antibacterial, and anticancer properties (Bao et al., 2021; Gao et al., 2013). Historically, jujube fruit was consumed both as dried and fresh fruit, and fresh jujube is warmly welcomed because of its high nutrient content and more delicious taste. However, the development of fresh jujube has been neglected due to its limited shelf life. In the past 30 years, increased cultivation and supply of fresh jujube have been witnessed owing to the improvement of jujube fruit preservation technology and marketing logistics.

6.1.12 *Antioxidant and antimicrobial properties of Jujube* (Ziziphus jujuba) *of Muscat*

Jujube has higher levels of phenolics and ascorbic acid than other common fruit varieties, according to research findings. Furthermore, to reveal the variations in the antioxidant and free radical scavenging capacities of various jujube cultivars, evaluations of the antioxidant capacity of jujube extract were conducted based on a decrease in power utilizing scavenging methods. Rutin, however, was not a metric used in these evaluations. Anagnostopoulou et al. (2006) discovered that the physicochemical characteristics and antioxidant capacity of jujube are cultivar-driven.

Research has demonstrated the therapeutic benefits of *Z. jujuba* as a medicinal plant, and investigations on the antimicrobial activities of jujube essential oils have reported positive antibacterial effects (Zhang et al., 2010). Jujube essential oils, thus, possess antibacterial qualities, and the plant's many sections, including its leaves, offer medicinal qualities as well. In fact, the antibacterial and antifungal activities of jujube fruit are what give it its biological characteristics (Fu et al., 2011).

6.2 *Conclusion*

Fruits are important to the food industry, but they are also important in the pharmaceutical and medical sectors. Many fruits are recognized for their medicinal uses due to their special and valuable properties. However, many herbs are either neglected or insufficiently researched. To fully realize the potential of these fruits and herbs, more investigation is required. The exploration of value addition in muscat fruits promises to significantly enhance health and quality of life.

Fruits are important to the food industry, but they are also crucial to the pharmaceutical and medical sectors. Many fruits are recognized for their medicinal uses due to their unique and valuable properties. However, many herbs are either neglected or insufficiently researched. To fully realize the potential of these fruits and herbs, further investigation is required.

References

Anagnostopoulou, M. A., Kefalas, P., Papageorgiou, V. P., Assimepoulou, A. N., & Boskou, D. (2006). Radical scavenging activity of various extracts and fractions of sweet range peel (Citrus sinensis). *Food Chemistry*, 94, 19–25.

Çalişkan, O., & Polat, A. A. (2011). Phytochemical and antioxidant properties of selected fig (*Ficus carica* L.) accessions from the eastern Mediterranean region of Turkey. *Scientia Horticulturae*, 128(2011), 473–478.

El-Hamamsy, S., & El-Khamissi, H. (2020). Phytochemicals, antioxidant activity and identification of phenolic compounds by HPLC of pomegranate (Punica granatum L.) peel extracts. *Journal of Agricultural Chemistry and Biotechnology*, 11(4), 79–84.

Fu, L., Xu, B. T., Xu, X. R., Gan, R. Y., Zhang, Y., Xia, E. Q., & Li, H. B. (2011). Antioxidant capacities and total phenolic contents of 62 fruits. *Food Chemistry*, 129, 345–350.

Ghazanfar, S. A., & Al Sabahi, A. M. A. (1993). Medicinal plants of Northern and Central Oman. *Economic Botany*, 47(1), 89–98.

Ghnimi, S., Umer, S., Karim, A., & Kamal-Eldin, A. (2017). Date fruit (*Phoenix dactylifera* L.): An underutilized food seeking industrial valorization. *NFS Journal*, 6, 1–10.

Hemalatha, M., Thirumalia, T., Saranya, R., Elumalai, E. K., & David, E. (2013). A review on antimicrobial efficacy of some traditional medicinal plants in Tamilnadu. *Journal of Acute Disease*, 2(2), 99–105.

Hong, Y. J., Tomas-Barberan, F. A., Kader, A. A., & Mitchell, A. E. (2006). The flavonoid glycosides and procyanidin composition of Deglet Nour dates (*Phoenix dactylifera*). *Journal of Agricultural and Food Chemistry*, 54, 2405–2411.

Juices, F. (2018). National Nutrient Database for Standard Reference Release Legacy April, 2018 Full Report (All Nutrients) 09087, Dates, United States Department of Agriculture, Agricultural Research Service, USDA Food Composition Databases. *Deglet Nour a.* 2018, 4–9.

Koehn, F. E. (2009). Drug discovery from natural products. *Nature Reviews Drug Discovery*, 8, 678.

Lulekal, E., Asfaw, Z., Kalbessa, E., & Van Damme, P. (2013). Ethnomedicinal study of plants used for human ailments in Ankober District, North Shewa Zone, Amhara Region, Ethiopia. *Journal of Ethnobiology and Ethnomedicine*, 9, 63.

Nakilcioğlu, E., & Hışıl, Y. (2013). Research on the phenolic compounds in sarilop (Ficus carica L.) fig variety. *GIDA*, 38(5), 267–274.

Panza, O., Conte, A., & Del Nobile, M. A. (2021). Pomegranate by-products as natural preservative to prolong the shelf life of breaded cod stick. *Molecules*, 26(8), 2385. https://doi.org/10.3390/molecules26082385

Selim, S., El Alfy, S., Al-Ruwaili, M., Abdo, A., & Al Jaouni, S. (2012). Susceptibility of imipenem-resistant *Pseudomonas aeruginosa* to flavonoid glycosides of date palm (*Phoenix dactylifera* L.) tamar growing in Al Madinah, Saudi Arabia. *African Journal of Biotechnology*, 11, 416–422.

Taleb, H., Maddocks, S. E., Morris, R. K., & Kanekanian, A. D. (2016). Chemical characterisation and the anti-inflammatory, anti-angiogenic and antibacterial properties of date fruit (*Phoenix dactylifera* L.). *Journal of Ethnopharmacology*, 194, 457–468.

stenberg-Hägg, J., Zagrobelny, M., & Bak, S. (2013). Plant defense against insect herbivores. *International Journal of Molecular Sciences*, 14(5), 10242–10297.

Vinson, J. A. (1999). The functional food properties of figs. *Cereal Foods World*, 44, 82–87.

Zhang, H., Jiang, L., Ye, S., Ye, Y. B., & Ren, F. Z. (2010). Systematic evaluation of antioxidant capacities of the ethanolic extract of different tissues of jujube (Ziziphus jujuba Mill.) from China. *Food and Chemical Toxicology*, 48, 1461–1465.

chapter seven

Development of nutraceuticals from medicinal fruits of Bahrain

K. Yogeshwari, Shivani Chauhan, Gurdeep Rattu, Divya Chauhan, and Sachin Kumar

7.1 Introduction

The surge in industrial activities, environmental pollution, and sedentary habits has contributed to a global rise in degenerative conditions and chronic health issues, which has led to a dramatic increase in healthcare requirements and, additionally, their costs. Often, conventional techniques come with adverse side effects, which have led the scientific community and medical professionals to focus their attention on the ancient practices of herbal medicines for the treatment and prevention of a variety of medical disorders. These natural remedies are being researched as potential replacements for conventional pharmaceutical medications. The surge in the trend of nutrients as medicines in the pharmaceutical industries is of great significance and draws the interest of researchers and scientists toward the potential benefits. This has led to the adoption of interdisciplinary approaches for designing and developing various herbal products optimized for their applications (Rajat, Manisha, Robin, & Sunil, 2012).

The term 'nutraceutical' was coined by Stefen Defelice, chairman of the Foundation for Innovation in Medicine. It is a hybrid of two terms: 'nutrition,' meaning nourishment of the body, and 'pharmaceutical,' which signifies a drug used for health benefits or disease treatment. Therefore, nutraceuticals can be defined as products that offer health benefits and aid in disease prevention and treatment (Donno, Mellano, Cerutti, & Beccaro, 2018). These nutraceuticals can be categorized into several groups based on their origin and chemical constituents, as illustrated in Figure 7.1 and Table 7.1 (Rajat et al., 2012).

In recent years, research has proven that many ailments are strongly linked to lifestyle choices and diet. A healthy diet with a mindful lifestyle has been proven to prevent chronic conditions such as obesity, thyroid issues, diabetes, cholesterol problems, cancer, anxiety, etc. (Rajat et al., 2012). Nutraceutical products have been well studied and have shown

 DOI: 10.1201/9781003452614-7

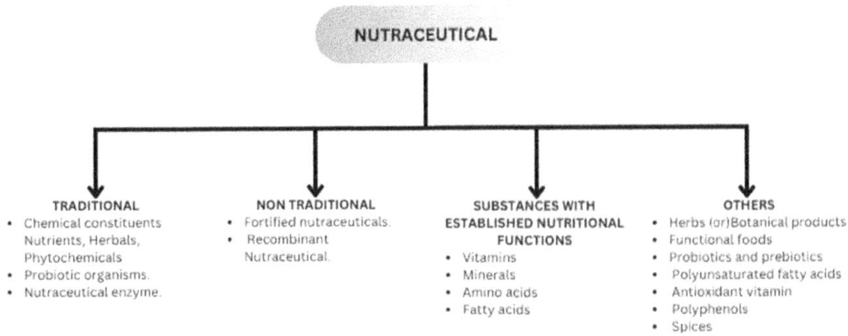

Figure 7.1 Classification of nutraceuticals.

promising benefits in improving overall health, effectively preventing lifestyle disorders, and extending lifespan. A brief overview of these benefits is illustrated in Figure 7.2.

The notion of nutraceuticals as food for both nutrition and medicinal purposes is nearly 3,000 years old. It has been practiced in many ancient cultures. From ancient civilizations, Bahraini healers have relied on their traditional healing practices with indigenous herbal plants to treat various diseases and maintain a healthy life. Their traditional healing techniques have been passed down through generations orally, making them an integral part of Bahrainis' lives (Al-Attar & Ahmad, 2016). People in India and China have consumed various natural foods such as turmeric, neem, black sesame, ginseng, and many more as part of their diet regularly; these herbs are now well studied and are proven to have promising health benefits. Similarly, countries like Germany, France, and England have recognized the importance of diet in preventing chronic diseases. They believe a healthy diet is more crucial than both exercise and hereditary factors for preventing chronic diseases such as cataracts, cardiovascular diseases, diabetes, gallstones, inflammatory diseases, neurodegenerative diseases (Alzheimer's/Parkinson's/Huntington's disease), obesity, and several types of cancers, ultimately achieving good health with increased life expectancy.

The nutraceutical industry has now become a multi-billion-dollar industry in the global marketplace. According to a recent analysis, the nutraceutical market size is estimated to be USD 488.41 billion in 2024. It is expected to reach USD 626.02 billion by 2029, growing at a CAGR of 5.09% during the forecast period. This growth is driven by a shift towards preventive healthcare, a rise in chronic diseases, and growing consumer interest in healthy eating. Furthermore, technological advancements, such as the adoption of AI for personalized nutrition solutions, are expected to positively influence consumer demand further (Grand View Research, 2025).

Table 7.1 Nutraceutical types and their examples

Nutraceutical type	Definition	Examples
Traditional nutraceutical	It is a naturally occurring food that does not undergo any manual changes. E.g., fruits and vegetables	Lycopene from tomato, guava, and papaya are potential anti-oxidant agents that protect against cancer (leukemia, cervical, and prostate)
Non-traditional nutraceutical	External supplement of nutrients to boost overall nutritional content	Beta carotene from carrots protects the cornea against UV light and has anti-oxidant, anti-carcinogenic, and anti-cancer properties
Fortified nutraceuticals	Fortification of food with the addition of micronutrients such as vitamins and minerals	Fortified milk with cholecalciferol for Vitamin D deficiency
Recombinant nutraceuticals	Application of biotechnology and genetic engineering in the production of energy-providing foods	Gold kiwi fruit is rich in ascorbic acid, carotenoids, lutein, and zeaxanthin
Functional foods	Food sources of necessary nutrients are required for growth, maintenance, and development	Cereals (rice, corn, wheat, etc.) are known to reduce the risk of coronary heart disease, lower blood pressure, and tumor incidence
Dietary supplements	Food products made of supplementary dietary ingredients are consumed as a remedy for deficiencies or diseases. It is not consumed as a conventional food regularly	Dietary supplements include vitamins, minerals, amino acids, etc. It aids in overall health improvement, fitness, performance improvement, and bodybuilding

7.2 *Fruits as nutraceuticals*

From ancient times, fruits have been an integral part of the human diet, many of which are used as medicines. Fruits are traditional nutraceuticals that promise consumers benefits beyond essential nutrition; they comprise vitamins, minerals, carbohydrates, protein, fibers, phenols, essential amino acids, antioxidants, and several other phytochemicals that offer

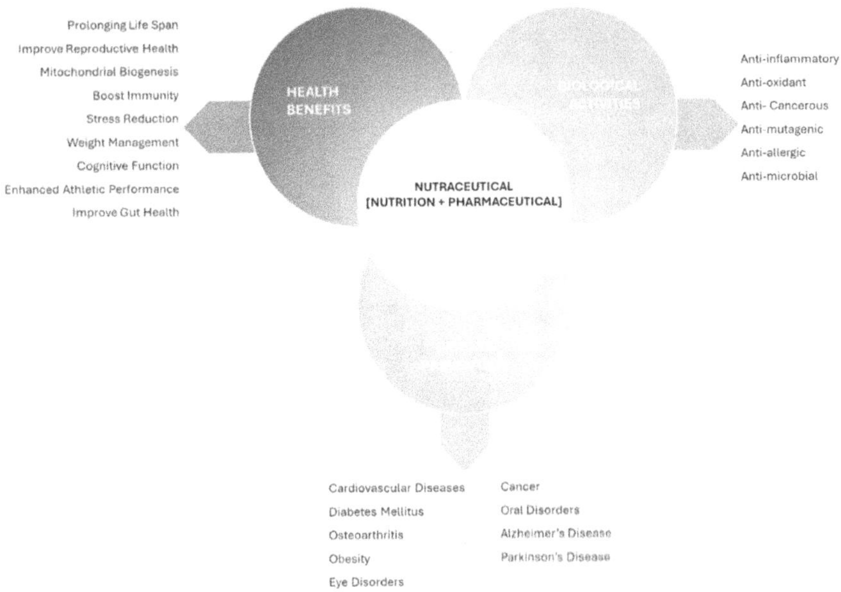

Figure 7.2 General overview of benefits of the nutraceuticals.

Table 7.2 Different colored fruits provide different phytochemicals

Color group	Phytochemicals
Red	Lycopene, anthocyanin
Yellow/Orange	Carotene, flavonoids
Green	Lutrin, glucosinolates
Blue/Purple	Anthocyanins, phenolics
White/Tan/Brown	Allyl sulfides, polyphenols

potential health benefits. Different fruits differ in their composition and color, as described in Table 7.2, based on which they offer a variety of phytochemicals and nutrients beneficial for overall health (Dolkar, Bakshi, Wali, Sharma, & Shah, 2017). A list of the most prominent biological activities of nutraceuticals present in fruits is outlined in Table 7.3.

7.3 Medicinal fruits of Bahrain

Bahrain is an island kingdom characterized by high temperatures, limited arable land, water scarcity, and increasing groundwater salinity. Situated in the Persian Gulf, it is well known for its tropical climate, which consists of various habitats such as conglomerates, sand dunes, salty plains (sabkhas), and mangrove areas, making it home to 323 plant species

Table 7.3 List of biological activities of nutraceuticals present in fruits with examples

Biological activity	Definition	Nutraceuticals	Examples
Anti-oxidant	Group of compounds that prevent damage caused by free radicals to the cells. They reduce oxidative stress and help in maintaining the cellular integrity.	Phenolic acids, flavonoids	Flavonols, catechins, anthocyanins, isoflavonoids, gallic acid, vanillic, ellagic, etc.
Anti-inflammatory	Inhibit or reduce inflammation by blocking the body's inflammatory signals or chemicals	Phenolic acids, flavonoids	Iso-vitexin, iso-orientin, 11-deoxycucurbita cinI-2-*O*-β-D
Anti-microbial	Group of compounds that kill or inhibit the growth of bacteria, fungi, and viruses	Alkaloids, cucurbitacin, terpenes, terpenoids, phenols	Choline, colocynthosides A, curcubitacin B, thymol, linalool, geraniol, linalyl acetate, carvacrol, and menthol
Anti-tumor	Chemicals or compounds that target and eliminate tumor cells or prevent their growth and spread	Cucurbitacin, saponin, flavonoids	Iso-orientin 3′-methylether, colocynthosides A, curcubitacin B
Anti-diarrheal	Compounds are used in reducing the severity of diarrhea	Phenolic compounds, flavonoids, glycosides, steroids	Isoflavonoids, cucurbitacin E 2-*O*-D-glucopyrano side.
Anti-allergic	Class of medication used to alleviate symptoms associated with allergic reactions	Glycosides	Cucurbitacin E 2-*O*-D-glucopyrano side, cucurbitacin E
Anti-mutagenic	Class of chemicals that protect cells from mutagenesis	Polyphenolic compounds	Quercetin, catechins

belonging to 196 genera and 53 families (Abbas, El-Oqlah, & Mahasneh, 1992). Notably, 52 of these species, classified within 49 genera and 29 families, are known to have medicinal properties (Rizk, 1986).

A study conducted on the use of herbal medicines in Bahrain demonstrated that 95.3% of respondents rely on traditional medicines for treating non-serious illnesses on the island (Alkhuzai, Al-Laith, & Mandeel, 2010). Similarly, another study also showed that 70% of the plants used in these traditional practices were native species (Alkhuzai, Mandeel, & Al-Laith, 2010). While several fruits show potential for nutraceutical use, most were introduced later with technological advancements and urbanization. There is no clear evidence that these introduced fruits are used in traditional Bahraini medicine. However, research identifies the following fruits as having a role in Bahraini medicinal practices, including *Citrullus colocynthis*, *Prosopis farcta*, *Ochradenus baccatus* Del., *Zizphus spina-christi* L., and *Datura fastusa* (Abbas et al., 1992; Alkhuzai, Mandeel, et al., 2010).

7.3.1 Citrullus colocynthis

General Description: It is a perennial plant belonging to the *Cucurbitaceae* family. The plant is widely distributed in tropical and subtropical regions of the world. It is mainly found in the deserts of Arabia and the Sahara, and in the southern part of Asia (Bhasin, Singh, & Garg, 2020; Rao & Poonia, 2023). Common fruit names include bitter apple, wild gourd, and desert gourd. In Bahrain, it is known as 'Murrah' or 'Handhal'. The leaves, fruits, and seeds of the plant have been boiled for ages to prepare a decoction by Bahrainis to treat ailments such as jaundice and hydragogue (Abbas et al., 1992; Alkhuzai, Al-Laith, et al., 2010).

Nutritional Profile: The fruits of *C. colocynthis* are high in moisture content, reaching over 90% of their total weight. They contain 70% unsaturated fatty acids, 51% polyunsaturated fatty acids, and proteins rich in essential amino acids such as methionine, arginine, tryptophan, etc. The fruit also contains several micronutrients essential for consumers. The complete nutrient profile of the fruit and seed is presented in Table 7.4.

Bioactive Compounds and Their Medical Significance: Cucurbitacin is the primary bioactive compound found in *C. colocynthis*. It plays a crucial role in the development of drugs for cancer therapy due to its cytotoxic behavior. Cucurbitacin exists in 12 different forms, with cucurbitacin E being the most abundant in the fruit, followed by cucurbitacin A, B, C, and D. The fruit also contains phenolic acids and flavonoids, including iso-vitexin, iso-orientin, and iso-orientin 3′-methyl ether. These compounds contribute to its anti-tumor, anti-inflammatory, and anti-diarrheal activities. Additionally, the choline alkaloid present in the fruit exhibits antimicrobial and antihelmintic activity, while saponins contribute to its anti-cancer properties (Rao & Poonia, 2023).

Table 7.4 Nutritional profile of C. *colocynthis* fruit

Micro nutrients	Amount (%)	Macro nutrients	Amount
Calcium	174.00	Carbohydrates	10.88%
Phosphorous	690.00	Fat	3.15%
Potassium	50.00		30.0% (mature fruit)
Sodium	8.50	Protein	24.37% (immature fruit)
Iron	16.80	Moisture	90%
Zinc	2.63		
Copper	6.90	**Fatty acid profile**	**Amount (%)**
Magnesium	390.00	Total saturated fatty acid	18.03
		Total unsaturated fatty acid	81.74
		Monounsaturated fatty acid	27.43
		Polyunsaturated fatty acid	54.25

Current Usage and Availability: The plant is extensively found in the deserts of Arabia and the Sahara, as well as the southern part of Asia, mainly India, Pakistan, and the southern islands (Coffey, Simmons, Shepard, Tadmor, & Levi, 2015). Traditionally, different plant parts have been used to treat many ailments, such as joint pain, leprosy, bronchitis, cancer, and mastitis (Heydari, Homayouni, Hashempur, & Shams, 2016). Additionally, the plant has medicinal benefits for gut-related disorders such as dysentery and indigestion. Apart from medicinal applications, the plant is also used in various industries, including the cosmetic, fragrance, and food industries (Rao & Poonia, 2023).

7.3.2 Prosopis farcta

General Description: It is a legume from the *Fabaceae* family (formerly *Leguminosae*). This prickly shrub or subshrub thrives in hot and dry conditions. It is widely distributed across arid, semi-arid, tropical, and subtropical regions. *Prosopis farcta* is resistant to heat, drought, salinity, and alkalinity, which allows it to flourish in the Bahrain archipelago. Locally known as Yanboot in Bahrain, its roots and fruits are used in traditional medicine to treat dysentery and spasms (Abbas et al., 1992; Zhong et al., 2022).

Nutritional Profile: *Prosopis* comprises various macronutrients (carbohydrates, fat, protein) and micronutrients (minerals). The carbohydrates most commonly present are galactose and mannose, with minor amounts of glucose and arabinose. The fatty acid profile shows abundant palmitic acid, followed by oleic acid, and a low amount of caprylic acid, capric acid, lauric acid, and pentadecanoic acid. Minerals present include potassium, magnesium, calcium, and phosphorus, which are particularly abundant, while copper, zinc, iron, and manganese are present in trace quantities (Zhong et al., 2022).

Bioactive Compounds and Their Medical Significance: *P. farcta* contains various plant chemicals. These include flavonoids like catechin and quercetin. These antioxidants help protect the body from harmful free radicals. Additionally, research suggests quercetin has anti-inflammatory, antioxidant, and antiplatelet properties, and 1-nicotine present in the plant exhibits antimicrobial properties (Jahromi, Etemadfard, & Zebarjad, 2018).

Current Usage and Availability: *P. farcta*, though considered a common weed in some countries, holds medicinal significance. It is widely distributed throughout North Africa and Asia. Getting rid of the plant is particularly difficult, as it requires the complete removal of roots from the soil. Despite its weedy property, the plant has established uses in traditional medicines. It is a known and available treatment for angina (chest pain). It is extensively used in therapies for cardiovascular disorders and in relieving heart or chest pain (Dheeb, Hashim, Hameed, & Hasan, 2020).

7.3.3 Ochradenus baccatus *Delile*

General Description: It is an annual or perennial herb belonging to the 'Resedaceae' family. It is geographically distributed in Bahrain, Saudi Arabia, and tropical Southwest Asian countries. In Bahraini folk medicine, the herb is used as an anti-inflammatory and antibacterial agent. Locally known as 'Gurdee', its fruits and leaves are commonly used to treat stomach pain (Kumar, Patyar, Alsaadi, & Patyar, 2024).

Bioactive Compounds and Their Medical Significance: The phytochemical examination of *O. baccatus* detected the presence of alkaloids, coumarins, saponins, fatty acids, and steroidal compounds. Specific components identified under these categories are quercetin, kaempferol glycosides, and phenolic compounds such as rutin, luteolin, afzelin, and astragalin. Notably, the plant also showed high levels of glucosinolates, which are another class of beneficial compounds. Though the plant is extensively used in folk medicine, there is a lack of scientific evidence to support its medicinal properties (Kumar et al., 2024).

Current Usage and Availability: *O. baccatus* is a tropical plant cultivated in dry, rocky areas. It is extensively distributed in the Arabian Peninsula, parts of Africa, and the Indian subcontinent. The plant is broadly used in folk medicines for treating headaches, fever, and sexual disorders, and as an antibiotic. Other applications include its use as a cooling agent and deodorant. The plant is also used as an ornamental plant in households and gardens and additionally serves as an essential source of food for certain desert animals (Mohammed, Ragab, & Khan, 2023).

7.3.4 Zizphus spina-christi

General Description: It is a tropical tree belonging to the *Rhamnaceae* family and is known for its rich source of diverse chemical compounds. These compounds contribute to the tree's antibacterial, anti-inflammatory, antioxidant, and various other medicinal properties (Abdulrahman et al., 2022). In Bahrain, it is called 'Sedr,' and its common English names include Christ's thorn and Syrian Christ-thorn. Local people traditionally use its fruits and leaves to treat respiratory problems and headaches. Medicinal preparations involve direct consumption or topical application (Abbas et al., 1992).

Nutritional Profile: The fruits of *Ziziphus spina* are typically low in moisture, averaging around 13.9%. This low moisture content makes them ideal for long-term storage with a reduced risk of microbial contamination or fermentation. However, these fruits are rich in carbohydrates, constituting about 73.8%, making them a good energy source. The fruit contains 1.1% lipid (fat), 6.0% crude protein, and 5.1% total ash. While total ash indicates mineral content, this ratio suggests the fruits are either deficient or moderately adequate in minerals (El Maaiden, El Kharrassi, Moustaid, Essamadi, & Nasser, 2019).

Bioactive Compounds and Their Medical Significance: The fruit of *Ziziphus spina* is rich in various phytochemicals with potential medicinal benefits. It exhibits a diverse range of polyphenols, including flavonoids like catechin, rutin, isoquercitrin, quercetin, and kaempferol. These flavonoids play a crucial role in antioxidant activity, acting as free radical scavengers. Additionally, the fruit contains cyclopeptides (Mauritine F, Sanjonine F, and Spinanine-A) and saponins (Christinin A and Jujubogenin-3-O-(di-deoxyhexosyl) hexoside). Notably, the polyphenolic compounds in *Ziziphus spina* exhibit anti-mutagenic and anti-carcinogenic properties, potentially providing health benefits for humans (Abdulrahman et al., 2022; El Maaiden et al., 2019).

Current Usage and Availability: The species is native to Africa, specifically the Sahara and Sahelian zones. Cultivation of this plant extends to the oases of the Sahara, Tunisia, Algeria, Niger, East Africa, Sinai, Iran, and North India, among other places (Orwa, Mutua, Kindt, Jamnadass, & Anthony 2009). The plant serves various purposes; the seed of *Ziziphus spina-christi* is used as feed for animals (El Maaiden et al., 2019). They are also used as alternatives for food preservatives due to their antimicrobial properties. Furthermore, the analysis revealed that the fruits of *Ziziphus spina-christi* are rich in ascorbic acid, which helps in disease prevention and improves fertility (Daja, Mohammed, Igein, & Yusuf, 2019).

7.3.5 Datura fastuosa

General Description: It is a perennial sub-herb belonging to the *Solanaceae* family and is native to tropical and subtropical regions of Asia and Africa. Now widely cultivated in warmer areas, it is valued for its ornamental uses and medicinal properties in traditional practices (Soosairaj, Maheswari, & Balaguru, 2010). However, the plant is also well-known for its toxic effects at high doses. Locally known as 'Benj' in Bahrain, the various parts of the plant are used medicinally; the fruit of *Datura fastuosa* is a 5 cm long, 4-valved capsule, densely thorny and walnut-sized. It is historically used in Bahrain to treat asthma and arthritis (Abbas et al., 1992; Alkhuzai, Al-Laith, et al., 2010).

Nutritional Profile: The preliminary phytochemical investigation was performed on a methanolic and hydroalcoholic extract of *Datura fastuosa*; dried seeds of the plant revealed the presence of alkaloids, tannins, cardiac glycosides, flavonoids, carbohydrates, amino acids, and phenolic compounds. The proximate analysis showed that the plant seeds were rich in macronutrients and micronutrients (Al-Snafi, 2017). The nutritional profile of *D. fastuosa* seeds is illustrated in Table 7.5.

Bioactive Compounds and Their Medical Significance: The fruit of *Datura fastuosa* possesses a variety of bioactive compounds with potential medicinal applications. Preliminary phytochemical analysis revealed the presence of tropane alkaloids, including hyoscyamine, hyoscine (scopolamine), and atropine. These alkaloids are most concentrated in the seeds, reaching up to 0.426%. Further analysis of the seeds identified additional compounds like isofraxidin, scopatone, daturadiol, 1,4-benzenediol, arenarine D, vanillin, N-trans-feruloyl-tyramine, scopoletin, β-sitosterol, and hyoscyamilactol. Studies have also found phenolic and flavonoid content in *D. fastuosa*, with the highest total phenolics (TPC) concentration of 268.6 μg of gallic acid equivalence/mg of dry plant material found in methanolic seed extracts. In comparison, the fruit pulp exhibited the highest total phenolic content (TPC) of 8.84 μg of quercetin equivalence/mg dry plant material (Al-Snafi, 2017).

Alongside, the *Datura fastuosa* fruit extracts exhibit various pharmacological effects, including antimicrobial, insecticidal, repellent, and

Table 7.5 Nutritional profile of *D. fastuosa* seeds

Chemical constituent	Amount (%)
Carbohydrates	51.22
Fat	14.72
Protein	20.73
Moisture	4.63
Crude fiber	17.35

organophosphate protective properties. However, it is crucial to note that the presence of potent alkaloids like hyoscyamine, hyoscine, and atropine can be highly toxic in high doses. Due to this inherent toxicity, specific dosages of any plant part from *D. fastuosa* should only be used under the guidance of a qualified professional, since overconsumption of the plant extracts can lead to lethal consequences (Al-Snafi, 2017).

Current Usage and Availability: The plant is found in tropical and temperate regions and is cultivated worldwide for its ornamental and medicinal uses. It is commonly available in tropical parts of India, Southeast Asia, Africa, and Central and South America (eFlora of India, 2025). The plant is known for its medicinal properties due to its tropane alkaloids. It is used in Ayurvedic medicine as a narcotic, antispasmodic, antitussive, bronchodilator, anti-asthmatic, and hallucinogenic. It has also been used to treat diarrhea, skin diseases, epilepsy, hysteria, rheumatic pains, hemorrhoids, painful menstruation, skin ulcers, wounds, and burns (Al-Snafi, 2017).

7.4 Challenges and future perspectives

Several obstacles must be overcome before Bahraini fruit-based nutraceuticals can be fully realized. For one, there is still a need to develop standardized objective scientific evaluations and evidence from robust human trial data on their disease-preventative properties and safety. Moreover, given the interplay among different bioactive compounds in the studied fruits, additional research should focus on comparing their preventative properties and discovering relevant novel biomarkers.

Additionally, most Bahraini fruits are underexplored, which limits their potential not only as food sources but also as economically and socioculturally valuable resources. Therefore, it is required to maximize research on their phytochemical profiles, antioxidant activities, and potential as novel functional foods and nutraceuticals. Overall, overcoming these challenges and solidifying the evidence on phytochemical composition, nutritional value, and scientific basis for traditional uses will allow Bahraini fruit nutraceuticals to be fully appreciated.

7.5 Conclusion

Though fruits are a significant part of the food industry, they also play a crucial role in the pharmaceutical and medicinal industries. It is well known that several fruits are used for medicinal purposes due to their unique and invaluable properties; however, countless herbs remain either underutilized or inadequately studied. Further research is necessary to uncover the full potential of these fruits and herbs.

Exploring nutraceuticals from Bahraini fruits holds great promise for significantly enhancing health and quality of life. While challenges remain, they can be addressed through focused research on phytochemical composition, nutritional value, and the scientific validation of traditional uses. By doing so, the potential benefits of Bahraini fruits can be realized as natural resources for effective and innovative disease prevention strategies. Furthermore, this research can contribute to both the economic and cultural development of the region. The integration of empirical healing knowledge with scientific understanding paves the way for a healthier future.

Acknowledgments

This research article was supported by The Gandhi Institute of Technology and Management, Visakhapatnam, Reva University, University of Delhi South Campus, and National Horticulture Research and Development Foundation (NHRDF), India.

Conflicts of interest

The authors declare no conflict of interest, financial or otherwise.

Ethical approval

Ethics approval was not required for this work.

References

Abbas, J. A., El-Oqlah, A. A., & Mahasneh, A. M. (1992). Herbal plants in the traditional medicine of Bahrain. *Economic Botany, 46*, 158–163.

Abdulrahman, M. D., Zakariya, A. M., Hama, H. A., Hamad, S. W., Al-Rawi, S. S., Bradosty, S. W., & Ibrahim, A. H. (2022). Ethnopharmacology, biological evaluation, and chemical composition of Ziziphus spina-christi (L.) Desf.: A review. *Advances in Pharmacological Pharmaceutical Sciences, 2022*, 4495688.

Al-Attar, M., & Ahmad, J. (2016). History of medicine in Bahrain territory. *Journal of Research on History of Medicine, 5*(3), 171–180.

Al-Snafi, A. E. (2017). Medical importance of Datura fastuosa (syn: Datura metel) and Datura stramonium-A review. *IOSR Journal of Pharmacy, 7*(2), 43–58.

Alkhuzai, J., Al-Laith, A., & Mandeel, Q. (2010). Knowledge, attitudes, and practices regarding medicinal plants used by the indigenous people in Bahrain. *Arab Gulf Journal of Scientific Research, 28*, 105–117.

Alkhuzai, J., Mandeel, Q., & Al-Laith, A. (2010). Folk medicine and the use of herbal plants in the Kingdom of Bahrain. *Arab Gulf Journal of Scientific Research, 28*(2), 91–104.

Bhasin, A., Singh, S., & Garg, R. (2020). Nutritional and medical importance of Citrullus colocynthis-A review. *Plant Archives, 20*(2), 3400–3406.

Coffey, J. L., Simmons, A. M., Shepard, B. M., Tadmor, Y., & Levi, A. (2015). Potential sources of whitefly (Hemiptera: Aleyrodidae) resistance in desert watermelon (Citrullus colocynthis) germplasm. *HortScience, 50*(1), 13–17.

Daja, A., Mohammed, Z. K., Igein, T. C., & Yusuf, A. D. (2019). A preliminary study on the nutraceutical potentials of aqueous fruit pulp extracts of Zizipus spina christii, Zizipus mauritiana and Chrysophyllum albidum. *Research Journal of Science, 19*(1), 35–46.

Dheeb, B. I., Hashim, S. S., Hameed, R. K., & Hasan, S. W. (2020). Phytochemistry and biological activity of Prosopis fracta. *Journal of Biotechnology Research Center, 14*(1), 25–33.

Dolkar, D., Bakshi, P., Wali, V., Sharma, V., & Shah, R. A. (2017). Fruits as nutraceuticals. *Ecology Environment and Conservation, 23*, S113–S118.

Donno, D., Mellano, M. G., Cerutti, A. K., & Beccaro, G. L. (2018). Nutraceuticals in alternative and underutilized fruits as functional food ingredients: Ancient species for new health needs. In: Holban, A. M., & Grumezescu, A. M. (Eds.), *Alternative and Replacement Foods* (pp. 261–282): Elsevier.

eFlora of India. (n.d.). Datura metel (L.). Retrieved May 8, 2025, from https://efloraofindia.com/efi/datura-metel/

El Maaiden, E., El Kharrassi, Y., Moustaid, K., Essamadi, A. K., & Nasser, B. (2019). Comparative study of phytochemical profile between Ziziphus spina christi and Ziziphus lotus from Morocco. *Journal of Food Measurement Characterization, 13*(1), 121–130.

Grand View Research. (2025). Nutraceuticals market size, share & trends analysis report by product (dietary supplements, functional foods, functional beverages), by ingredient, by application, by region, and segment forecasts, 2024–2030. Grand View Research Report. Retrieved from https://www.grandviewresearch.com/industry-analysis/nutraceuticals-market

Heydari, M., Homayouni, K., Hashempur, M. H., & Shams, M. (2016). Topical C itrullus colocynthis (bitter apple) extract oil in painful diabetic neuropathy: A double-blind randomized placebo-controlled clinical trial. *Journal of Diabetes, 8*(2), 246–252.

Jahromi, M. A. F., Etemadfard, H., & Zebarjad, Z. (2018). Antimicrobial and antioxidant characteristics of volatile components and ethanolic fruit extract of Prosopis farcta (Bank & Soland.). *Trends in Pharmaceutical Sciences, 4*(3), 177–186.

Kumar, D. L., Patyar, R. R., Alsaadi, M. A. T. A., & Patyar, S. (2024). A review on phytochemical and pharmacological aspects of Ochradenus baccatus. *BIO Web of Conferences, 86*, 01019.

Mohammed, H. A., Ragab, E. A., & Khan, R. A. (2023). Taily weed (Ochradenus baccatus Delile) as a middle Eastern Herb: Chemical profile, biological activities, and traditional uses. In: Rajendram, R., Preedy, V., & Patel, V. (Eds.), *Ancient and Traditional Foods, Plants, Herbs and Spices Used in the Middle East* (pp. 293–306): CRC Press.

Orwa C, Mutua A, Kindt R, Jamnadass R, S Anthony. (2009). Agroforestree database: A tree reference and selection guide version 4.0. Retrieved from https://www.worldagroforestry.org/output/agroforestree-database.

Rajat, S., Manisha, S., Robin, S., & Sunil, K. (2012). Nutraceuticals: A review. *International Research Journal of Pharmacy, 3*(4), 95–99.

Rao, V., & Poonia, A. (2023). Citrullus colocynthis (bitter apple): Bioactive compounds, nutritional profile, nutraceutical properties and potential food applications: A review. *Food Production, Processing and Nutrition, 5*(1), 4.

Rizk, A.-F. M. (1986). *The Phytochemistry of the Flora of Qatar*: Scientific and Applied Research Centre, University of Qatar.

Soosairaj, S., Maheswari, A., & Balaguru, B. (2010). Micromorphological assessment in Datura species of Tamil Nadu, India. *Journal of Economic Taxonomic Botany, 34*(4), 873–879.

Zhong, J., Lu, P., Wu, H., Liu, Z., Sharifi-Rad, J., Setzer, W. N., & Suleria, H. A. (2022). Current insights into phytochemistry, nutritional, and pharmacological properties of Prosopis plants. *Evidence-Based Complementary Alternative Medicine, 2022*, 2218029.

chapter eight

Bioactive properties of processed products of medicinal fruits from Kuwait

Rohini Dhenge, Arjun Mohanakumar, Massimiliano Rinaldi, and Tommaso Ganino

8.1 Introduction

Medicinal fruits have been integral to traditional medicine for centuries, offering a wealth of bioactive compounds with potential therapeutic benefits. In Kuwait, the arid climate and unique biodiversity have led to the cultivation and use of several indigenous medicinal fruits renowned for their health-promoting properties. The processing of these fruits into various forms—such as powders, extracts, and infusions—can influence their bioactive content and efficacy. Understanding these properties is essential for leveraging their full potential in nutraceuticals and functional foods. Fruits such as *Ziziphus spina-christi* (Sidr), *Phoenix dactylifera* (Date palm), and *Punica granatum* (Pomegranate) are prominent in Kuwaiti traditional medicine. Each of these fruits is valued not only for its nutritional benefits but also for its significant pharmacological properties. In general, these fruits not only provide medicinal benefits but also play an important role in maintaining overall health and well-being in Kuwait, reflecting the region's deep-rooted connection to natural, plant-based remedies. Research on Kuwaiti medicinal fruits emphasizes their abundance of bioactive compounds and significant health benefits. Processing these fruits has been shown to enhance their therapeutic properties, making them valuable in nutrition and healthcare. Continued research focuses on optimizing processing techniques and further exploring their bioactive profiles to maximize their potential applications.

8.1.1 Ziziphus spina-christi *(sidr)*

Sidr (*Ziziphus spina-christi*), commonly known as "Christ's thorn," holds a prominent place in Kuwaiti traditional medicine due to its

 DOI: 10.1201/9781003452614-8

anti-inflammatory, antimicrobial, and antioxidant properties. The fruit, leaves, and seeds of Sidr are rich in flavonoids and phenolic compounds, which contribute significantly to its therapeutic efficacy (Elanany and Salem, 2024). *Ziziphus spina-christi* (L.) Willd, a member of the Rhamnaceae family, is an armed shrub or evergreen tree known by several common Arabic names, including Sidr and Nabeq. This plant is native to a vast area of Africa and is widely distributed across the tropical and subtropical regions of the world. *Ziziphus* plants, in general, have a long history of use in traditional medicine for the treatment of various ailments. Among these, *Ziziphus spina-christi* is particularly noted for its diverse biologically active ingredients, including alkaloids, and sterols such as β-sitosterol, flavonoids, triterpenoids, sapogenins, and saponins. These components endow the plant with potent antibacterial, antioxidant, antimicrobial, and anti-inflammatory properties. Given its rich array of active compounds and the ease with which it can be cultivated, *Ziziphus spina-christi* is considered a valuable tree with significant potential for the development of new therapeutic drugs in the future (Hussein et al., 2019).

Atwaa et al. (2022) explored the impact of sidr fruit pulp (*SFP*) on the functional and nutritional properties of fermented camel milk and its effects on diabetic rats. The study involved supplementing camel milk with 5%, 10%, and 15% *SFP*, ultimately selecting the 15% concentration for its superior properties. In a trial with 32 diabetic rats, the group fed fermented camel milk with 15% *SFP* showed improved health outcomes compared to those receiving only camel milk or no treatment, highlighting the potential benefits of *SFP* in managing diabetes. The study found that supplementing camel milk with sidr fruit pulp (*SFP*) significantly enhanced its total solids, protein, ash, fiber, viscosity, phenolic content, and antioxidant activity, with these improvements increasing alongside the *SFP* ratio. Specifically, the total phenolic content (TPC), total flavonoids (TF), and DPPH% antioxidant activity of the camel milk were 0.85 mg/g, 0.08 mg/g FW, and 6.34%, respectively, which were consistent with the higher values found in fresh *SFP* itself (10.40 mg/g, 1.24 mg/g FW, and 62.02%). These results align with previous research indicating that sidr fruit is rich in bioactive compounds like phenolics and flavonoids.

Elanany and Salem (2024) provided a detailed analysis of the bioactive compounds in *Ziziphus spina-christi* (Sidr), highlighting significant differences between Sidr pulp powder and fresh pulp. The total phenol content in Sidr pulp powder was found to be 179.53 mg GAE/100 g DW, while total flavonoids were 411.72 mg QE/100 g DW. Both the pulp powder and water extract of Sidr exhibited high antioxidant activity, with the pulp showing 96.02% DPPH inhibition, the highest among the samples tested. The study further examined the phenolic compounds using HPLC, revealing notable variations between the fresh pulp and the pulp powder. Pyrogallol was present at 62.77 mg/g in the fresh pulp but only 13.28 mg/g in the pulp powder, indicating its higher concentration in the fresh state. Similarly, 3 oH-Tyrosol and Catechol were slightly more concentrated in the pulp powder (2.99 and 1.42 mg/g) compared to the fresh pulp (2.91 and 1.35 mg/g, respectively). Notably, four Amino Benzoic Acid was found in the fresh pulp at 10.34 mg/g but was absent in the pulp powder. Catechin, a major bioflavonoid, was abundant in both forms, though slightly reduced in the fresh pulp (153.93 mg/g) compared to the pulp powder (179.63 mg/g). Chlorogenic Acid was detected only in the fresh pulp (35.95 mg/g), suggesting its specific enrichment in the pulp. Other compounds such as P-oH Benzoic Acid and Benzoic Acid were significantly higher in the fresh pulp (40.32 and 30.09 mg/g) compared to the pulp powder (4.06 and 4.19 mg/g). Caffeic Acid also followed this trend, being more concentrated in the fresh pulp (6.81 mg/g) than in the pulp powder (0.86 mg/g). Caffeine, present in the fresh pulp at 36.30 mg/g but absent in the pulp powder, further supports the idea that some compounds are reduced or lost during processing. The fresh pulp was also the sole source of Ellagic Acid (457.32 mg/g) and Salicylic Acid (116.87 mg/g), known for their potent antioxidant and anti-inflammatory properties. Coumarin was found at significantly higher levels in the fresh pulp (86.51 mg/g) compared to the pulp powder (1.05 mg/g). Ferulic acid, however, was absent in both forms.

Flavonoid analysis revealed that several key flavonoids are present only in the fresh pulp, such as Apigenin 6-arabinose 8-galactose (88.40 mg/g), Apigenin 6-rhamnose 8-glucose (38.15 mg/g), and Kaempferol 3-2-P-Coumaryl Glucose (50.03 mg/g). Rutin, a major antioxidant, was significantly higher in the fresh pulp (281.31 mg/g) compared to the pulp powder (2.54 mg/g). Rosmarinic Acid also showed a higher concentration in the fresh pulp (65.82 mg/g) than in the pulp powder (0.59 mg/g). Quercetin was found only in the fresh pulp at 62.99 mg/g, while Quercitrin was exclusive to the pulp powder (7.10 mg/g). Hesperidin was present only in the pulp powder (14.45 mg/g), and Naringenin was more concentrated in the pulp powder (2.65 mg/g) compared to the fresh pulp (4.10 mg/g). Kaempferol was significantly higher in the fresh pulp (18.03 mg/g) than in the pulp powder (0.17 mg/g). Apigenin, an anti-inflammatory agent, was also found in both forms, with slightly higher levels in the fresh pulp (3.11 mg/g) compared to the pulp powder (0.09 mg/g). Overall, these results underscore that the fresh pulp of *Ziziphus spina-christi* retains a richer profile of bioactive compounds compared to its dried counterpart, highlighting its potential as a potent source for nutraceuticals and functional foods.

The study by Elanany and Salem (2024) underscores the potential of *Ziziphus spina-christi* (Sidr) as a valuable source of nutrients and health-promoting compounds, advocating for its use in food products like cakes and crackers. In cakes, the antioxidant activity and bioactive compound levels increased significantly with higher Sidr pulp concentrations.

Specifically, 15% Sidr pulp resulted in an antioxidant activity of 81.64% DPPH, with total phenols and flavonoids at 559.55 mg GAE/100 g DW and 621.99 mg QE/100 g DW. At 20%, these values peaked at 84.46% DPPH, 640.94 mg GAE/100 g DW, and 890.29 mg QE/100 g DW, respectively. In crackers, a 15% Sidr pulp concentration led to an antioxidant activity of 58.75% DPPH, with total phenols and flavonoids at 279.20 mg GAE/100 g DW and 142.79 mg QE/100 g DW. Increasing to 20% raised these metrics to 60.31% DPPH, 304.53 mg GAE/100 g DW, and 148.38 mg QE/100 g DW, while the highest concentration of 30% resulted in 78.53% DPPH, 317.33 mg GAE/100 g DW, and 150.63 mg QE/100 g DW. Overall, the incorporation of Sidr pulp powder significantly enhances the antioxidant capacity and bioactive compound levels in both cakes and crackers, highlighting its potential as a functional food ingredient.

In a study conducted by Tawfek et al. (2022), nabq fruit pulp (*Ziziphus spina-christi*) was used as a substitute in low-fat ice milk, with varying levels of cream replacement at 75% and 100%. The results indicated that 100% replacement with nabq fruit pulp yielded a total phenol content of 27.13 mg/100 g, while 75% replacement resulted in 24.56 mg/100 g. The antioxidant activity was also measured, showing 30.66% activity at 100% replacement and 22.38% at 75% replacement. Lower levels of replacement, such as 25% and 50%, demonstrated reduced phenol content and antioxidant activity. Additionally, fresh nabq fruit pulp, on a fresh weight basis, exhibited an antioxidant activity of 68% and a total phenol content of 173 mg GAE/100 g. This study highlights the potential of nabq fruit pulp as a valuable ingredient in enhancing the nutritional profile of low-fat ice milk products.

In a study by EH et al. (2021), dried nabq fruit (*Ziziphus spina-christi*) was utilized to develop low-fat yogurt. The study reported that yogurt containing 4% dried nabq fruit exhibited a total phenol content (TPC) of 80.30 mg/100 g. Additionally, another variant with 2% dried nabq fruit showed a TPC of 62.80 mg/100 g. This research underscores the potential of dried nabq fruit as a functional ingredient in enhancing the phenolic content and nutritional value of low-fat yogurt.

8.1.2 Phoenix dactylifera *(dates)*

Phoenix dactylifera Linn, commonly known as the date palm, belongs to the Palmaceae family and thrives in arid regions globally (Nasir et al., 2015). This versatile plant, which includes over 2,000 varieties such as Medjool and Deglet Noor, is prized for its fruit and has significant commercial and health value (Siddiq and Greiby, 2013). Dates are rich in dietary fiber, antioxidants, and essential nutrients like potassium, iron, and calcium, contributing to their growing use in functional foods and bioactive ingredients (Mousavi et al., 2014; Ayad et al., 2020). Research highlights their anti-diabetic, anti-inflammatory, and anti-cancer properties, reinforcing their role in health maintenance and disease prevention (Al-Farsi et al., 2005). In regions such as the Sahara, dates are integral to both diet and traditional medicine (Al-Turki et al., 2010). Clinically, their potential benefits on pregnancy outcomes are being explored (Mirza et al., 2018). With about 70% of their carbohydrate content in the form of easily absorbed sugars, dates provide substantial energy, making them a nutritious option for all age groups and individuals with chronic conditions (Kamal-Eldin et al., 2012; Ahmed et al., 2013).

Fresh dates, due to their high moisture content, must be dried to prolong shelf life and enhance consumer appeal.

- Dried dates are preferred for their sweeter taste and easier storage. Optimal drying conditions involve temperatures of 60°C–70°C and an air velocity of 2 m/s to preserve color and texture (Ilce Gabriela et al., 2017). Various drying methods, including microwave and conventional techniques, are also assessed for their effects on antioxidant and nutritional properties (Elsharawy et al., 2019; Echegaray et al., 2020).
- Date paste, a product of minced, pitted dates, is rich in sugars, fiber, and phenolic compounds, with added date seeds increasing their fiber content (Karizaki, 2017). Date jam and jelly are prepared from dates at different stages, with the jam cooked to a TSS of 68°Brix and the jelly to 65°Brix, incorporating additional ingredients like citric acid and pectin (Vijayanand and Kulkarni, 2012; Echegaray et al., 2023).
- Date butter, made from "tamar" stage dates, is boiled to a TSS of 74°–75° Brix and adjusted to a pH of 4.5–4.7 (Echegaray et al., 2020).
- Date candy, chutney, and relish, prepared from various date stages and processed with different spices and preservatives, offer diverse uses (Elsharawy et al., 2019; Benyahia-Krid et al., 2021; Salomón-Torres et al., 2021).
- Pickled dates, dried and preserved with spices, vary in storage longevity depending on the date stage (Alyahya et al., 2022).
- Date seed oil, extracted through milling and solvent processing, is used in soaps, while the deoiled cake serves as animal feed (Meghwar et al., 2021; Alyahya et al., 2022).

8.1.2.1 Date as a functional ingredient in food products

Aleid and Haddadin (2023) assessed the total polyphenol content and antioxidant activity of date fruit, date syrup, and date pits. The total polyphenol content was highest in date pits (2,954 ± 9.22 mg GAE/100 g), followed by date fruit (1,833.52 ± 8.5 mg GAE/100 g) and date syrup

(1,448.65±8.5 mg GAE/100 g). Antioxidant activity, measured by IC50, was strongest in date syrup (62.43±0.77 μg/mL), followed by date fruit (88.23±0.52 μg/mL) and date pits (280±6.54 μg/mL). This indicates that date pits have the highest polyphenol content, while date syrup has the most potent antioxidant activity. For example, Al-Farsi and Lee (2008) reported a higher phenolic content in seedless press cake (2.76 mg/g) and antioxidant activity than in date syrup (1.33 mg/g). The appearance of this date press cake is a powder with a light brown color due to the natural pigments from date flesh and seeds, mainly carotenoids and anthocyanins.

A study conducted by Djaoud et al. (2020) stated that date syrup (DS) and date powder (DP) were utilized in various concentrations, guided by a D-optimal mixture design, to formulate dairy desserts: syrup dairy dessert (SDD), mixture dairy dessert (MDD), and commercial dairy dessert (CDD). The addition of DS and DP to these dairy desserts (SDD and MDD) resulted in a significant increase in phenolic content ($p<0.05$) compared to the CDD, enhancing their antiradical and reducing power activities. Among the desserts, SDD exhibited the highest total phenolic content (TPC), with values recorded as 0.65±0.04 mg/g for SDD, 0.38±0.03 mg/g for MDD, and 0.27±0.00 mg/g for CDD. In terms of DPPH radical scavenging activity, the percentages were 4.00%±0.82% for SDD, 3.22%±0.44% for MDD, and 2.54%±0.46% for CDD. The reducing power was measured at 2.75±0.10 mg/g for SDD, 2.21±0.17 mg/g for MDD, and 0.17±0.16 mg/g for CDD. Notably, incorporating 16% DS in SDD and a combination of 14% DS and 2% DP in MDD significantly boosted the TPC to 0.65±0.04 and 0.38±0.03 mg/g, respectively, compared to 0.27±0.00 mg/g in the CDD. This demonstrates the positive impact of DS and DP on the nutritional and antioxidant properties of dairy desserts.

In a recent study, Shahein et al. (2022) investigated the effects of adding date syrup as a prebiotic and flavoring agent to probiotic fermented camel milk, assessing its physicochemical, phytochemical, microbiological, and sensory properties over 15 days of storage at 5°C. Date syrup was added at 6% and 8%, with non-flavored fermented camel milk serving as the control. The addition of date syrup significantly enhanced the total phenolic content (TP) and antioxidant activity (AO) of the fermented milk. Initially, TP levels were 1.68±0.12 mg/100 g in the control, but increased to 22.40±1.98 and 40.28±1.86 mg/100 g in the 6% and 8% date syrup samples, respectively. By day 15, TP in the control dropped to 0.94±0.11 mg/100 g, while the 6% and 8% samples maintained higher levels at 17.32±1.74 and 36.56±2.16 mg/100 g. Total flavonoid (TF) content and antioxidant activity followed similar trends. On day 1, TF was 0.54±0.02 mg/100 g in the control, rising to 3.42±0.94 and 4.86±0.90 mg/100 g in the 6% and 8% samples. By day 15, TF levels dropped in the control but remained higher in the date syrup samples. Antioxidant activity was initially 6.33%±1.0% in the control, increasing to 24.70%±2.18% and 29.84%±1.94% in the 6% and 8% samples, with slight decreases by day 15 but still highest in the 8% sample.

The fruit of the date palm is rich in polyphenols, which can be effectively utilized in developing functional food products, as evidenced by various studies. For instance, Gad et al. (2010) explored the impact of replacing part of the water in reconstituted skim milk powder with varying concentrations of date palm syrup for yogurt production. Their findings indicated that yogurt enriched with 10% date syrup exhibited a significant increase in total phenolic content, hydrogen-donating capacity, and ferric-reducing antioxidant power, thereby enhancing the yogurt's health-promoting properties. Almusallam et al. (2021) found that extracts from date palm spikelets of the Reziz and Khalas cultivars were rich in phenolics and flavonoids, exhibiting over 94% DPPH and ABTS radical scavenging activity. In a subsequent study, the researchers added this extract at 0.5% and 1% concentrations to cow's milk yogurts and assessed their properties during cold storage. The spikelet extract enhanced the yogurts' total polyphenol and flavonoid content, leading to increased antioxidant capacity (DPPH improved from 10.87% to 63.12%–75.33% and ABTS from 18.22% to 37.73%–42.81%) and reduced lipid oxidation, as indicated by lower TBARs values (from 2.50 to 1.13–1.25 mg MDA/kg) after 21 days.

Hassan et al. (2024) conducted a study to evaluate the phenolic and flavonoid content in date cupcakes with different concentrations of date pulp. The study reported the Total Phenolic Content (TPC) and Total Flavonoid Content (TFC) in these samples. For the cupcakes containing 20% date pulp, the TPC was found to be 97.66±9.56 mg GAE/100 mL, while the TFC was measured at 70.02±2.6 µg/mL. In contrast, for the cupcakes with 25% date pulp, the TPC increased to 98.94±8.64 mg GAE/100 mL, with a corresponding TFC of 71.5±3.4 µg/mL. This study highlights the impact of varying date pulp concentrations on the antioxidant properties of date cupcakes.

Abdel-Maksoud et al. (2022) found that incorporating date seed powder (DSP) into beef meatball samples at varying levels (4%–16%) significantly enhanced the content of polyphenol and flavonoid compounds, as well as other nutritional values, leading to a marked improvement in antioxidant properties. There was a highly significant increase ($p < 0.05$) in polyphenol and flavonoid levels and antioxidant activity (ORAC) in both uncooked and cooked beef meatballs as the DSP content increased. Specifically, the total polyphenols were recorded at 176.82, 351.13, 542.77, and 692.15 mg/100 g, total flavonoids at 22.11, 43.77, 64.90, and 85.81 mg/100 g, and ORAC values at 40.75, 79.75, 118.65, and 157.82 µmol trolox equivalents/g for the DSP levels of 4%, 8%, 12%, and 16%, respectively. These results were significantly higher compared to the control beef meatball samples, which had values of 1.66 mg/100 g for polyphenols, 0.39 mg/100 g for flavonoids, and 1.29 µmol Trolox equivalents/g for ORAC.

The study by Famuwagun and Gbadamosi (2021) obtained the DPPH radical scavenging activities of the bread samples, which varied from 32.05% to 61.63%. Bread sweetened with wet dates exhibited stronger DPPH scavenging activity than bread sweetened with dry dates, and both had greater activity than bread sweetened with granulated sugar. Additionally, bread enriched with date fruit had a higher phenolic content than bread with granulated sugar, likely due to the phenolics in the dates. Bread sweetened with wet dates had higher phenolic content compared to those with dry dates. Both 15% dry and wet date additions (1 mg GAE/g) showed similar but higher phenolic content than the 5% additions.

In the realm of baked goods, incorporating functional ingredients has been shown to boost the nutritional and health benefits of these products. A previous study demonstrated that substituting sucrose with date palm fruit in bread increased the dietary fiber content without significantly affecting the bread's nutritional quality (Nwanekezi et al., 2015). Similarly, muffins enriched with 2.5% and 5% date fruit fiber concentrates (DFC) had higher dietary fiber content, antiradical activity, and secondary oxidation inhibition, particularly in those baked at 165°C with 5% DFC (Mrabet et al., 2015). These findings underscore that the addition of functional ingredients like dates, which are high in dietary fiber and polyphenols, can enhance the functional and health benefits of baked goods, as noted in previous research (Mrabet et al., 2015). A traditional dish known as Idli, commonly made with sugar and other ingredients, is popular in India and Sri Lanka. However, the high sugar content in Idli can have negative health effects, such as increasing the risk of dental caries and diabetes. A study aimed at mitigating these risks substituted sugar with date syrup, paste, and chopped dates in Idli preparation (Manickavasagan et al., 2013). The results showed that Idli made with dates had superior sensory qualities and overall acceptability. Additionally, the incorporation of dates increased the total phenolic and vitamin C content compared to Idli made with sugar. Therefore, dates can be used to enhance the health benefits of traditional foods like Idli.

*8.1.3 Pomegranate (*Punica granatum*)*

The pomegranate (*Punica granatum L.*), a member of the Punicaceae family, is native to a region stretching from Iran to the Himalayas. Known by various names such as the Chinese apple, Apple of Carthage, and Apple with Many Seeds, pomegranates thrive in tropical and subtropical climates. They are widely cultivated in countries like Iran, India, Spain, and the USA, as well as in parts of the Near and Far East (Schubert et al., 1999). In global trade, India has achieved the highest growth rate in pomegranate exports to Kuwait, at 42.22%, a figure that is statistically significant at the 1% level. Conversely, the Netherlands recorded the lowest import growth rate at 3.54%. Kuwait also ranked second in the growth of total import value, with a rise of 61.25% (Aware et al., 2019). Major importers of pomegranates include the UAE, Oman, Saudi Arabia, the Netherlands, Qatar, Iran, Kuwait, Nepal, and Bangladesh. Historically, every part of the pomegranate tree—roots, bark, leaves, flowers, rind, and seeds—has been utilized in medicine due to its rich chemical content. The fruit's sweet varieties are known for their laxative properties, while those with a balance of sweetness and acidity are believed to soothe stomach inflammation and heart pain. Recent studies noted that pomegranates may also enhance the activity of enzymes that protect against cardiovascular risks (Delgado et al., 2020). Pomegranates are gaining recognition as a valuable crop because of their high polyphenol content, particularly ellagic acid, known for its potent antioxidant and anti-inflammatory effects (Aviram et al., 2000). Research highlights the potential benefits of pomegranate extracts in supporting cardiovascular health and cancer prevention. The fruit's chemopreventive properties are well-documented, addressing conditions such as colic, colitis, diarrhea, dysentery, leucorrhea, paralysis, and headaches (Sadeghi et al., 2009; Jurenka, 2008). In traditional medicine, particularly within Ayurvedic and Unani systems, pomegranates are valued for bioactive compounds like betulinic and ursolic acids, as well as alkaloids such as pseudopelletierine and pelletierine.

Commercial farming of pomegranates is expanding globally, driven by their health benefits, including antioxidant, antimutagenic, and antihypertensive properties. Pomegranates are also known for their ability to reduce liver damage. The anthocyanins and polyphenols in pomegranates help scavenge free radicals and enhance the body's antioxidant capacity (García-Pastor et al., 2020). Additionally, pomegranates possess anti-inflammatory and anti-atherosclerotic properties, with potential therapeutic effects against osteoarthritis, prostate cancer, cardiovascular disease, and HIV-I (Malik et al., 2005; Sumner et al., 2005). Pomegranate juice, regarded as one of nature's most potent antioxidants, strengthens the immune system, cools the body, and supports kidney, liver, and heart function (Alper et al., 2005; Nuncio-Jáuregui et al., 2015; Gil et al., 2000).

Pomegranates also show significant promise for processing into value-added products with extended shelf life. While large, well-colored fruits are highly preferred by consumers, lower-quality pomegranates with imperfections, such as sunburned husks or cracked skins, can still be utilized in the processing industry. Despite their high potential, industrial processing of pomegranates remains limited due to challenges in peeling and the lack of advanced processing technologies (Yadav et al., 2006; López-Rubira et al., 2005). Typically, pomegranates contain 48%–52% edible parts, with juice making up 78%–80% of the arils and seeds comprising

20%–22%. Juice extraction is a key method of value addition, producing a product celebrated for its flavor, fragrance, taste, and nutritional and medicinal benefits. This juice is further processed into various products, including squash, syrup, nectar, jelly, and concentrate, and is often used to impart color to other food items.

8.1.3.1 Pomegranate as a functional ingredient in food products

8.1.3.1.1 Frozen arils, appetized arils and arils preserved in vinegar The arils were frozen after being packed in polyethylene bags containing syrup with a concentration of 150°Brix, similar to the juice content of the arils. This method was used after preparing the arils in the same way as minimally processed arils. As the arils freeze, their juice mixes with the syrup. A similar freezing method involving sugar-coated arils was reported by Maestre et al. (2000), where the sugar coating turned red due to anthocyanin release during freezing and storage. To maintain their texture and avoid excessive softness, it is recommended that arils be consumed while still frozen. Appertized arils are made by immersing them in syrup with 150°Brix concentration, sealing them in metal tins, and heating them for 10 minutes to sterilize. After this stabilization process, the texture and taste of the arils were often compromised, becoming too soft and cooked. However, tins that underwent milder heat treatments retained a more desirable texture and flavor. Arils were preserved by placing them in 5% vinegar and sealing them in jars. This preservation method resulted in the arils turning a brownish color.

8.1.4 Pomegranate juice and jam

In a study by Mena et al. (2013), untreated pomegranate juice had an initial phenolic content of 1,350.34 mg GAE/L, which dropped to 967.10 mg GAE/L by day 35, a 39.6% reduction. However, phenolic compounds were somewhat resistant to high hydrostatic pressure (HHP) treatments, with content increasing by 3.38%–11.99% at 350 and 550 MPa pressures, respectively, compared to untreated samples. Polyphenol levels initially rose in treated juices over the first 3 days but started to decrease after 5 days, stabilizing after 10 days. In contrast, phenolic content in control juices continually decreased. HHP-treated samples showed a slight reduction in antioxidant capacity over time. DPPH scavenging activity, measured by IC50 values, ranged from 11 to 20 mg/mL on day 0. Juices treated at 450 and 550 MPa exhibited the highest antioxidant capacity (11–13 mg/mL), compared to the control (14 mg/mL) and 350 MPa treatment (15–18 mg/mL). Significant differences were found between control and HHP-treated samples. Antioxidant activity slightly increased during the first 5 days of storage, from 18 to 25 mg/mL, then decreased to 17 mg/mL by the end of the period. This increase in antioxidant activity might be due to the extraction of hydrolyzable tannins, ellagic acid, or anthocyanin polymers formed during storage (Pérez-Vicente et al., 2004).

The study Gaikwad et al. (2024) assessed how drying temperature and thickness affect the total phenolic content (TPC) and antioxidant capacity (AC) of foam mat-dried pomegranate juice (PJ). The drying conditions were categorized as T1 (70°C; 6 mm), T2 (70°C; 3 mm), T3 (60°C; 6 mm), T4 (60°C; 3 mm), T5 (50°C; 6 mm), and T6 (50°C; 3 mm). PJ had the highest TPC of 2,787 mg GAE/L, while foam mat-dried pomegranate juice (FPJP) had a TPC between 2,252 and 2,771 mg GAE/L, showing significant differences across treatments. TPC losses were greater at higher temperatures and thicknesses, with T1, T2, and T3 losing over 10%, whereas T4, T5, and T6 had a maximum loss of 2.14%. The total anthocyanin content (TAC) was 32.90 mg/100 mL in PJ, decreasing to between 28.83 and 21.43 mg/100 mL in FPJP, reflecting higher losses at elevated temperatures. TAC was highest for T6, with a 16.10% loss, whereas T1, T2, and T3 showed significant reductions. AC was lower in FPJP compared to PJ, which had an AC of 47.29 mg AAE/100 mL. Lower temperatures and thicknesses yielded higher AC. Increased thickness led to longer drying times and greater loss of bioactive compounds, reducing AC.

In a study by Esposto et al. (2021), eight 100% pomegranate juices (PJs) were examined, ICPJ1 (Internet, from concentrated pomegranate juice), INCPJ2 (Internet, not from concentrated pomegranate juice), SNCPJ3 (Store, not from concentrate pomegranate juice), INCPJ4 (Internet, not from concentrate pomegranate juice), SCPJ5 (Store, from concentrate pomegranate juice), SCPJ6 (Store, from concentrate pomegranate juice), ICPJ7 (Internet, from concentrate pomegranate juice), and INCPJ8 (Internet, not from concentrate pomegranate juice). CJs had higher total phenolic content (1,379–3,748 mg GAE/L) compared to NCJs (1,632–2,736 mg GAE/L), but lower anthocyanin levels (0.7–47 vs. 59.7–280.6 mg/L), likely due to degradation during concentration. NCJs had more volatile compounds (1,272–4,002 μg/L) than CJs (286.3–1,020 μg/L), with NCJs primarily containing alcohols (86%) and CJs containing more acids (44%) and furans (33%). The ethanol in NCJs suggests possible pre-treatment fermentation. Antioxidant activity strongly correlated with phenolic content (R values of 0.96 and 0.97 for DPPH and FRAP assays), with no significant differences between NCJs and CJs. No link was found between anthocyanin content and antioxidant activity, indicating that other compounds like hydrolyzable tannins are more influential.

In the study by Velotto et al. (2023), initial pomegranate juice (PJ) had a total polyphenol content (TPC) of 1,764.4 mg/L. Post-processing TPC varied by method: boule pilot plant jams (*B*) had TPCs between 1,990.2 and 2,387.9 mg/kg, microwaved jams (*M*) ranged from 1,335.9 to 1,812.7 mg/kg, and freeze-concentrated jams (*F*) exhibited the highest TPCs, from 4,218.8 to 5,555.5 mg/kg. Sonicated jams generally had lower TPC compared to unsonicated ones but were still higher than those processed by other methods. The initial PJ contained 110.04 mg/L of anthocyanins. After processing, boule-treated samples had anthocyanin levels from 42.0 to 111.2 mg/kg, microwaved jams ranged from 33.4 to 111.2 mg/kg, and freeze-concentrated jams had the highest levels, from 156.6 to 186.2 mg/kg. Antioxidant activity (AA) was highest in freeze-concentrated jams, with ORAC values ranging from 415 μM TE/100 g in MLM to 2,583.7 μM TE/100 g in FLM, and similar trends were observed in the DPPH assay. Ascorbic acid content in fresh PJ was 175.8 mg/L. Post-processing, freeze-concentrated jams had the highest ascorbic acid content, especially FHM with 267.3 mg/kg. Microwave-treated jams had the lowest content, with MLMU at 47.55 mg/kg. Boule-treated jams showed values from 64.7 mg/kg (BKU) to 84.4 mg/kg (BK), and microwaved jams ranged from 47.6 to 61.6 mg/kg in MLM. Sonicated samples generally had the lowest vitamin C content. Overall, freeze concentration methods were most effective in preserving polyphenols and ascorbic acid, while other methods led to notable losses.

In the study by Campos-Lozada et al. (2024), pomegranate juice was extracted using two methods: manual extraction with heat treatment (Man-P) and mechanical pressing with hyperbaric sanitization (Mech-Hyp). The Man-P method resulted in a higher phenolic content of 104,566 mg GAE/mL compared to 85.67 mg GAE/mL from Mech-Hyp. This difference was likely due to the higher water content in the Mech-Hyp samples, which may have diluted the phenolic compounds. For anthocyanins, the Man-P juice contained 15.140 mg/L of cyanidin 3-glucoside, while Mech-Hyp yielded 16.409 mg/L, indicating a significant impact of the extraction method on anthocyanin levels. Both extraction methods showed a high inhibition of the ABTS radical at 96.99%, surpassing previous studies, which reported lower values. In terms of DPPH radical scavenging, the Man-P method demonstrated a higher inhibition rate of 98.48%, compared to 87.93% for Mech-Hyp, indicating greater antioxidant activity in the manually pressed juice.

Pomegranate juice is concentrated and slowly heated over an extended period to produce a thick product called *Anar Rub*, which has good shelf stability. With a total soluble solids (TSS) content between 70% and 75%, it can be stored for up to a year and is commonly used as a jam. Maestre et al. (2000) reported that about 25% of the pigments in frozen Mollar pomegranate juice are lost during the jam-making process, which involves adding pectins, sucrose, and citric acid. The degradation of pigments continues during storage and is influenced more by temperature than by light, with a recommended storage temperature of 5°C. A visually appealing jelly can also be made from pomegranate juice, though around 50% of the anthocyanins in the juice are lost during the jelly-making process. Maestre et al. (2000) found that acidifying the juice significantly enhanced the jelly's color, both immediately and over time. However, color variations were noticed during storage, suggesting that factors other than pH also influenced the jelly's color stability.

In the study of Yun et al. (2023), eight organic acids were identified in pomegranate jam: tartaric, quinic, citric, fumaric, oxalic, malic, acetic, and lactic acids. The content of these organic acids was notably lower in jams processed at high storage temperatures (HSJ) compared to those processed at low temperatures (LSJ). This reduction in organic acids in HSJ is attributed to pyrolysis and decarboxylation reactions occurring during prolonged heating. Similarly, the total organic acid content decreased by 12% in "Earliblue" blueberry puree after heat treatment. For anthocyanins, the initial total anthocyanin content (TAC) was 95.4 mg/kg in LSJ and 73.2 mg/kg in HSJ. The extended heating during HSJ processing resulted in a significantly lower TAC compared to LSJ. However, High Pressure Processing (HPP) did not significantly impact TAC in either HSJ or LSJ. After 30 days of accelerated storage at 35°C, TAC levels were 8.6 mg/kg in HSJ, 9.5 mg/kg in LSJ-HPP, and 6.0 mg/kg in HSJ-HPP. The retention rate of TAC was higher in LSJ-HPP (10%) compared to HSJ-HPP (7%). The degradation of TAC followed a fractional conversion kinetic model, with the C_∞ value being higher for LSJ-HPP (18.01) than for HSJ-HPP (8.86). Overall, these results indicate that processing methods significantly affect the stability of organic acids and anthocyanins, with HPP demonstrating superior preservation of anthocyanins compared to thermal treatments.

8.1.5 Anardana and pomegranate syrup/grenadine

Anardana is made from the dried arils of sour, wild pomegranates and is commonly used as a souring agent in North Indian cuisine, often replacing tamarind or dried green mango (*amchur*) in dishes like curries, chutneys, and other culinary preparations. Known for its anti-inflammatory properties and high dietary fiber content, anardana is also used in the preparation of digestive candies and in traditional Ayurvedic and Unani medicine, particularly for treating ulcerative colitis. The improved method for processing anardana involves pre-cleaning, mechanized extraction of arils, solar or sun drying, and packaging. After treating the arils with 600 ppm sodium benzoate for 10 minutes, they are dried in a drier at 45°C for 48 hours to achieve a moisture content of 10%–12%. The final product has an appealing brown color and can be stored for extended periods in glass jars.

Pomegranate syrup, used in beverages and confectionery, is commercially known as grenadine. This syrup, made by mixing pomegranate juice with sugar, has a vivid purplish-red color, a pleasant taste, and contains about 60°Brix and 1.5% acidity from citric acid. It is preserved through pasteurization or sodium benzoate and stored in amber glass bottles at room temperature or in the refrigerator. Thakur et al. (2013) found that a combination of 40% arils and 60% syrup at 75°Brix was preferred for its sensory qualities and physicochemical properties.

8.1.6 *Pomegranate molasses*

Pomegranate molasses is a thick, tangy syrup made by simmering pomegranate juice until it reduces to a dark, sweet concentrate. The sweetness of the molasses comes from the natural sugars in the fruit (Yadav et al., 2006). Typically heated for about 45 minutes, it thickens without burning. Stored in an airtight container in the refrigerator, it lasts up to 3 months and is used in chutneys, curries, salad dressings, and as a meat glaze or tenderizer. Bekeir et al. (2024) stated that pomegranate molasses has garnered interest for its potential as a bioactive compound. Molecular docking studies showed that ellagic acid, gallic acid, punicalagin, and punicalin from pomegranate molasses bind more effectively to *E. coli* DNA gyrase B than novobiocin, with binding scores between –7.2 and –9.8 Kcal/mol compared to –6.30 Kcal/mol for novobiocin. These results suggest that pomegranate molasses could be a potent antibacterial agent and DNA gyrase inhibitor, offering a promising approach to combating antimicrobial resistance.

In Dargham et al. (2022), the total phenolic content of pomegranate molasses (PM) was 9.86–78.28 mg GAE/g for commercial samples and 21.19–44.32 mg GAE/g for homemade samples. The average content was 37.27 ± 10.79 mg GAE/g for homemade PM and 26.13 ± 14.32 mg GAE/g for commercial PM (*p*-value 0.087). The total flavonoid content ranged from 12.15 to 90.92 mg QE/g in commercial PM and 28.73 to 48.28 mg QE/g in homemade PM. Homemade PM had a significantly higher average flavonoid content of 40.09 ± 8.41 mg QE/g compared to 29.47 ± 17.61 mg QE/g in commercial PM (*p*-value 0.035). For antioxidant activity, the IC50 values for DPPH scavenging were 0.09–0.76 mg/mL for commercial PM and 0.09–0.26 mg/mL for homemade PM. The average IC50 for homemade PM was 0.15 ± 0.07 mg/mL, lower than 0.23 ± 0.15 mg/mL for commercial PM (*p*-value 0.231). Both types exhibited antioxidant activity similar to ascorbic acid (IC50 0.03 mg/mL).

8.1.7 *Pomegranate concentrate*

Pomegranate concentrate, made from pure pomegranate juice with no added sugars or preservatives, typically contains 65%–70% total soluble solids (TSS) and has a pH between 2.7 and 3.1. It is marketed globally for its health benefits. Maskan (2006) describes methods for preparing pomegranate concentrate, maintaining a Brix level of 60.5° using microwave, rotary vacuum, and atmospheric heating for 23, 108, and 190 minutes, respectively. Hulya Orak (2009) observed increases in reducing sugars, glucose, and fructose, as well as higher potassium and magnesium levels in the final concentrate.

In the study by Habib et al. (2023), the total phenolic content of the pomegranate juice concentrate (PJC) was 274.30±2.13 mg/100 g. The hydrolysis of glycosidic monomers and phenolic oligomers in PJC releases phenolic aglycons, which react with the Folin-Ciocalteu reagent to enhance total phenolic measurements. The total flavonoid content of the pomegranate fruit concentrate (PFC) was 82.58±0.75 mg/100 g. These findings align with some previous studies, though other reports have shown higher values, potentially due to differences in factors such as phenolic content, extraction methods, and sample conditions. The major phenolic compounds identified in PJC included 4-Hydroxy-3-Methoxybenzoic acid (909.72±28.66 mg/100 g), epicatechin (532.84±18.70 mg/100 g), catechin (310.51±1.07 mg/100 g), rutin (287.25±5.42 mg/100 g), ferulic acid (150.62±1.91 mg/100 g), P-coumaric acid (95.86±3.90 mg/100 g), and cinnamic acid (15.13±0.12 mg/100 g). Syringic acid and gallic acid were not detected. This study is the first to report the phenolic acid profile of PJC. Regarding antioxidant activity, the DPPH, scavenging activity of PJC was 19.17%±1.01%. In comparison, rutin showed a scavenging activity of 29.23%±0.41%, and vitamin C demonstrated 20.18%±0.32%. These results are consistent with prior studies and reflect the influence of various factors such as pomegranate variety, ripeness, growing conditions, and processing methods on antioxidant capacities.

8.1.8 *Pomegranate beverages*

Pomegranate juice can be processed into ready-to-serve (RTS) beverages, blends, and squash, which have a delicate flavor. RTS beverages and blends contain 15% juice, while squash contains 40% juice and is served by diluting one part squash with four parts water.

In another study, Kokkinomagoulos et al. (2020) examined three commercial yeasts (for wine, beer, and cider) to produce pomegranate alcoholic beverages (PABs) from the juice of the 'Wonderful' variety at 15°C and 25°C. The PABs contained ethanol ranging from 5.6% to 7.0% v/v, with glycerol content of 2.65–6.05 g/L and low volatile acidity. The process led to reductions in total flavonoids, phenolics, monomeric anthocyanins, and free radical-scavenging activity, yet 81%–91% of free radical activity, 29%–41% of phenolics, 24%–55% of flavonoids, and 66%–75% of anthocyanins were retained. The PABs contained 30 volatile compounds, including 15 esters, four organic acids, eight alcohols, and three terpenes.

Mantzourani et al. (2020) developed a functional pomegranate beverage by fermenting juice with *Lactobacillus paracasei* K5, immobilized on delignified wheat bran. The fermentation, conducted at pH levels of 3.3,

3.6, and 3.9 for 24 hours and then stored at 4°C for 4 weeks, effectively produced a low-ethanol (0.5%–1% v/v) and high-quality synbiotic pomegranate drink. The organic acids in fermented pomegranate juice increased from 1.67 g/L after 24 hours to 3.46 g/L by the 4th week. Ethanol concentration also rose from 0.5% to 1.0% (v/v) by the end of the 4th week. Adjusting the pH to 3.6 reduced total sugars by 34.5% to 61.1 g/L, while organic acids increased from 1.25 to 5.86 g/L, and ethanol increased from 0.5% to 1.0%. At pH 3.9, sugar reduction was 58.7% (50.1 g/L final concentration), organic acids increased from 1.17 to 4.86 g/L, and ethanol from 0.6% to 1.0%. These conditions kept alcohol content below 1%, making the beverage suitable for low or non-alcoholic consumption. The total phenolic content (TPC) of fermented pomegranate juice rose significantly from an initial 100 ± 10 mg GAE/100 mL to 111.6 ± 16.6 mg GAE/100 mL for non-fermented and 212.5 ± 29.9 mg GAE/100 mL for fermented samples, with a significant difference (t-test $t = -12.52$, $p < 0.05$). MF-ANOVA showed that the fermentation process ($F = 253.42$, $p < 0.05$) and storage time ($F = 11.41$, $p < 0.05$) were significant factors, while pH was not ($F = 1.12$, $p = 0.298$). TPC increased to 170–185 mg GAE/100 mL during the first 24 hours of fermentation and continued to increase over 3 and 4 weeks. Volatile compounds identified by headspace SPME GC/MS increased from 27 at the start of fermentation to 46 after 24 hours. The majority were carbonyl compounds, especially ketones. Notable exceptions included 2,2,5,5-tetramethyltetrahydrofuran (TMTHF), 2-methyl-1-penten-3-one, and 2,3-butanedione (diacetyl), which had not been previously found in fresh pomegranates or other fruits.

In the study by Rios-Corripio et al. (2020), the impact of various processing methods on pomegranate beverages was evaluated for antioxidant capacity (AC), total phenolic content (TPC), and total flavonoids (TFs). On Day 0, the highest AC was found in the HHP3 processed beverage (326.51 mg Trolox/100 mL), while the VAT pasteurized beverage had the lowest (296.48 mg Trolox/100 mL). By Day 42, VAT pasteurization led to the greatest reduction in AC, decreasing to 270.90 mg Trolox/100 mL, whereas HHP3 retained the highest AC at 324.88 mg Trolox/100 mL. Regarding TPC, HHP processed beverages consistently had higher values. Initially, HHP3 had the highest TPC (399.21 mg GA/100 mL) and maintained a high level (383.40 mg GA/100 mL) after 42 days. In contrast, the VAT pasteurized beverage decreased from 324.37 mg GA/100 mL to 270.90 mg GA/100 mL. For TFs, HHP3 showed the highest content both at Day 0 (121.54 mg Q/100 mL) and Day 42 (117.32 mg Q/100 mL). While HHP processing led to a slight increase in flavonoids, the change was not statistically significant. Overall, the study indicates that High Hydrostatic Pressure (HHP) methods are more effective than traditional pasteurization in preserving antioxidant capacity and phenolic content in pomegranate beverages.

8.1.9 Pomegranates confectionery

Pomegranate juice is processed by adding sugar and heating it until it reaches a thick consistency. Research indicates that about 50% of the anthocyanins in the juice are lost during jelly production, but acidifying the juice enhances the final product's color. *Anar Rub*, a pomegranate jam, is made by similarly cooking juice with sugar to a thick consistency. Pomegranate is also used to create jelly and hard candy. The study focused on developing jelly using a mix of cultivated and wild pomegranate fruit juices and evaluating its quality over a 6-month storage period. The optimal jelly formulation, based on sensory analysis, contained 55% mixed fruit juice (80% cultivated pomegranate and 20% wild pomegranate) and 45% sugar. The jelly maintained its quality in terms of chemical attributes and sensory characteristics—such as total soluble solids (TSS), titratable acidity, ascorbic acid, anthocyanins, pectin, total phenolics, color, texture, flavor, and overall acceptability—throughout the storage period. Storage conditions had an impact, with refrigerated storage showing slower changes compared to ambient conditions. After 6 months under refrigeration, TSS changed from 67.00°Brix to 67.36°Brix, titratable acidity from 0.75% to 0.71%, ascorbic acid from 9.18 to 7.35 mg/100 g, anthocyanins from 11.34 to 10.17 mg/100 g, and total phenols from 63.10 to 60.06 mg/100 g. Both polyethylene terephthalate (PET) and glass jars were effective packaging options, but glass jars showed a slight advantage in slowing down changes during refrigerated storage.

In the study by Cano-Lamadrid et al. (2020), varying gelatin percentages significantly affected the antioxidant capacity, total phenolic content (TPC), color coordinates, and texture of jellies. The study focused on different formulations, including pomegranate juice (PJ), sucrose (S), honey (H), and apple purée (AP), with the percentage of each component expressed by weight. The ABTS+ values ranged from 2.46 mg/100 g fresh weight (fw) in PJAPS-20 to 3.69 mg/100 g fw in PJAPS-13. Gelatin concentration notably influenced these values, with the highest ABTS+ observed in samples with 13% or 17% gelatin. DPPH values ranged from 10.6 mg/100 g fw in PJAPS-20 to 14.9 mg/100 g fw in PJS-20, with formulation and gelatin concentration being significant factors. FRAP values varied from 1.41 mg/100 g fw in PJAPH-20 to 4.59 mg/100 g fw in PJS-13, showing a similar trend to DPPH. TPC values ranged from 72.0 to 159 mg/100 g fw, with specific values being: PJH-20 (74.1 mg/100 g fw), PJS-20 (124 mg/100 g fw), PJAPH-20 (107 mg/100 g fw), PJAPS-20 (139 mg/100 g fw), PJH-20c (133 mg/100 g fw), PJS-20c (115 mg/100 g fw), PJAPH-20c (107 mg/100 g fw), and PJAPS-20c (160 mg/100 g fw). Overall, higher gelatin concentrations generally enhanced antioxidant capacity and TPC in the jellies.

Devhare et al. (2021) also developed a pomegranate juice-based *"chikki"* (a type of nut brittle) containing 20% juice, which maintained its quality for 3 months when packaged in LDPE bags. Furthermore, microencapsulated pomegranate peel extract (MEPPE) was successfully incorporated into chewing gum by Palabiyik et al. (2018), demonstrating its potential as a carrier for bioactive compounds. Other scientists have explored using pomegranate seed and peel extracts and oils in various confectionery products. For example, Lashkari et al. (2020) developed a whey-less fruit feta cheese that included 20% pomegranate juice, which showed improved antioxidant properties and oxidative stability, although it had a softer texture compared to the control.

8.1.10 Pomegranates wine

Pomegranate juice can be fermented into wine using starter wine yeast. The juice is sweetened to a Brix level of 22°–23° before fermentation, which is allowed to proceed until the desired alcohol content is achieved. Pomegranate wine typically contains 9%–10.1% alcohol and has a pH of around 4.0, with fermentation taking no longer than 7 days. The wine is clarified using bentonite treatment or centrifugation and aged in a manner similar to red grape wine.

In the study of Mena et al. (2012) it was observed that pomegranate wines showed key findings, highlighting changes in organic acids and antioxidant properties during winemaking. Citric acid levels were stable across wines, with concentrations of 1.29 g/100 mL in *Wonderful*, 0.64 g/100 mL in *Coupage*, and 0.28 g/100 mL in *Mollar de Elche*. Malic acid decreased from 0.30 g/100 mL in the juices to 0.10 g/100 mL during fermentation and then stabilized. Tartaric acid remained largely unchanged. Acetic acid increased during fermentation, reaching 0.008 g/100 mL in *Wonderful*, 0.005 g/100 mL in *Coupage*, and a smaller amount in *Mollar de Elche*. *Coupage* had the highest acetic acid levels between days 5 and 10 but saw reductions thereafter. *Wonderful* and *Coupage* retained some acetic acid after racking, while *Mollar de Elche* lost most.

Anthocyanins were significantly degraded, with cyanidin 3-glucoside decreasing by 76%–82%, delphinidin-3,5-diglucoside and cyanidin-3,5-diglucoside by 30%–49%, and pelargonidin glycosides reduced to negligible levels by day 2. Total phenolic content (TPC) ranged from 390 mg GAE/100 mL in *Wonderful* to 288 mg GAE/100 mL in *Mollar de Elche*, with post-winemaking reductions of 7% for *Wonderful*, 24% for *Coupage*, and 30% for *Mollar de Elche*. Antioxidant activity was highest in *Wonderful* wine, with reductions of 6%–30% for ABTS+ and 16%–39% for DPPH. *Wonderful* showed the smallest decrease in antioxidant activity, while *Mollar de Elche* had the largest decline. Despite these reductions, pomegranate wines retained high antioxidant values, with strong correlations between TPC, anthocyanins, ellagic acid, and antioxidant capacity.

8.1.11 Pomegranate powder and rind powder

Pomegranate arils are dried and milled to create pomegranate powder, a low-moisture product. Ellagic acid is the key active compound in pomegranate powder, and freeze-dried varieties rich in ellagic acid have multiple uses. This powder can be encapsulated for regular doses. Husain et al. (2004) proposed grade standards for pomegranate powder to bring it under AGMARK certification. The powder is used in various applications, including the production of "*yanggaeng*" (a traditional Korean dessert), bread, jellies, and other products.

Pomegranate rind has historically been used to produce ink in medieval Europe and is still utilized as a dye for textiles and specialty craft inks (Stover and Mercure, 2007). The peel, which makes up about 20%–50% of the fruit's total weight, is rich in essential minerals like potassium, calcium, phosphorus, magnesium, and sodium, as well as complex polysaccharides and a variety of bioactive compounds such as phenolics, flavonoids, proanthocyanidins, and ellagitannins like punicalagins and its isomers. Additionally, it contains smaller amounts of punicalin, gallic acid, ellagic acid, and ellagic acid glycosides. Due to its free radical-scavenging and antioxidant properties, pomegranate peel extract shows potential as a natural additive and functional ingredient when incorporated into real food products (Jalal et al., 2018). The primary polyphenol components of pomegranate peel powder include catechins, gallic acid, caffeic acid, ellagic acid, p-coumaric acid, and resorcinol (Rowayshed et al., 2013). Its high phenolic content and antioxidant activity, along with its crude fiber content, offer several health benefits, including improved glucose tolerance and insulin response, reduced LDL cholesterol levels, alleviated hyperlipidemia and hypertension, enhanced gastrointestinal health, and potential cancer prevention, particularly against colon cancer (Ranjitha et al., 2018). In the study by Ranjitha et al. (2018), the analysis of pomegranate peel powder revealed a total phenol content of 18.75 mg GAE/g and an antioxidant activity of 59.64%.

In the study of Giri et al. (2024) the impact of pomegranate peel powder (PPP) and chemical preservatives on the bioactive compounds and in-vitro antioxidant activity of muffins was evaluated. Notable improvements were observed with the inclusion of 8% PPP. Specifically, the total phenolic content increased significantly from 0.92 mg GAE/100 g in the control muffins to 12.5 mg GAE/100 g with 8% PPP, representing approximately a 12-fold enhancement. Similarly, total tannins rose from 0.2 mg GAE/100 g in the control to 8.27 mg GAE/100 g with 8% PPP. The in-vitro antioxidant activity, measured by the DPPH assay, improved from 6.97% in the control muffins to 29.34% with 8% PPP, reflecting a roughly six-fold increase. The FRAP assay showed a rise from 0.497 mg AAE/100 g in the control muffins to 2.934 mg AAE/100 g with 8% PPP. In contrast, muffins containing 0.1% potassium sorbate (PS) and 0.5% calcium propionate (CP), which are synthetic preservatives, did not exhibit significant differences in bioactive compounds or antioxidant activity compared to the control. For the 0.1% PS muffins, the total phenolic content was 0.87 mg GAE/100 g, total tannins were 0.3 mg GAE/100 g, DPPH antioxidant activity was 5.54%, and FRAP antioxidant activity was 0.544 mg AAE/100 g. The 0.5% CP muffins had total phenols of 0.83 mg GAE/100 g, total tannins of 0.4 mg GAE/100 g, DPPH antioxidant activity of 5.47%, and FRAP antioxidant activity of 0.399 mg AAE/100 g. Overall, the study highlighted that the addition of 8% PPP significantly boosted the bioactive compounds and antioxidant properties of muffins, while synthetic preservatives had minimal impact.

In Mohapatra et al. (2023), the impact of blending pomegranate peel powder with various spices on antioxidant activity and total phenolic content during storage was analyzed. Initially, the highest antioxidant activity, measured by DPPH inhibition, was in T2 (90 g pomegranate peel powder + 10 g turmeric powder) at 43.08%, followed by T5 (90 g pomegranate peel powder + 10 g onion powder) at 41.28%, and T6 (90 g pomegranate peel powder) at 36.87%. After 40 days, T5 retained the highest activity at 35.72%, with T3 (90 g pomegranate peel powder + 10 g ginger powder) at 34.37% and T2 at 34.09%. At 0 days, T2 had the highest phenolic content at 75.85 mg GAE/g, while T6 had the lowest at 66.97 mg GAE/g. After 40 days, T2 continued to lead with the highest phenolic content, followed by T5, T3, and T1 (90 g pomegranate peel powder + 10 g black pepper powder), with T6 having the lowest at 52.21 mg GAE/g. Blending with turmeric or onion powder maintained higher antioxidant activity and phenolic content during storage compared to other blends and the control.

The peel also has notable wound-healing properties (Chidambara Murthy et al., 2004). Its antioxidant-rich composition helps protect against bacteria and infections and is beneficial in skincare products, where it helps maintain skin pH balance, hydrate, and protect against pollutants. Detanninated peel powder, which separates hydrolysable tannins/ellagitannins while preserving nutritional components, can be used as a novel cattle feed supplement (Kushwaha et al., 2013). The incorporation of pomegranate peel powder meal (PPPM) in diets for broiler chickens enhanced feed conversion ratios, protein efficiency ratios, and various growth parameters, while also improving enzyme activity and nutrient digestibility. Additionally, it reduced serum aspartate aminotransferase levels at concentrations of 4–8 g/kg feed (Akuru et al., 2021). In sheep feeding, dried pomegranate peels used at a 1% level promoted optimal growth performance and reduced feed costs, offering a more economically efficient option (Omer et al., 2019). The antibacterial and antioxidant properties of meat burgers incorporating pomegranate rind powder extract (PRPE) improved due to its high phenolic content, which reduced lipid oxidation and showed promise as a natural alternative to synthetic antioxidants in high-fat meat products (Shahamirian et al., 2019). This powder also served as a cost-effective antioxidant for enhancing the quality of raw ground pork and fresh-cut fruit salads, as well as in value-added products like bread and cookies (Qin et al., 2013; Lacivita et al., 2021; Palak et al., 2020).

8.1.12 Pomegranate seed oil

Pomegranate seeds, a by-product of juice production, are rich in lipids, proteins, sugars, and essential minerals, with higher mineral content than the juice itself, except for potassium (Dadashi et al., 2013). They contain various nutraceuticals, including sterols, γ-tocopherol, punicic acid, hydroxyl benzoic acids (Liu et al., 2009; Rowayshed et al., 2013), and phenyl aliphatic glycosides like phenethyl rutinoside (Wang et al., 2004). Extracts from pomegranate seeds have demonstrated anti-diarrheal and antioxidant properties (Rowayshed et al., 2013; Singh et al., 2002). The oil is extracted by cold-pressing the seeds, which still retain 12%–20% oil after juice extraction. This oil is a significant source of conjugated octadecatrienoic fatty acids, notably punicic acid, along with other beneficial fatty acids like catalytic acid and alpha-eleostearic acid. Pomegranate seed oil is also rich in ω-3, ω-6, and ω-9 fatty acids, which are important for nutrition, heart disease prevention, cancer protection, and dissolving saturated fats (Rowayshed et al., 2013). It may also reduce leukotriene production from arachidonic acid, which is linked to asthma, skin inflammation, and cardiovascular diseases (Adhami and Mukhtar, 2006; Huxley and Neil, 2003). The oil's antioxidant power is lower than the peel but higher than the juice. Cold-pressing yields 13%–15% crude oil from dry seed material, with polyunsaturated fatty acids making up 87.82% of the oil (Kola et al., 2021). Encapsulation techniques using lipid-based carriers help protect the oil's sensitive components and maintain its nutritional and functional properties, with commercialization also extending to nanoemulsion forms.

Pomegranate peel serves as a functional ingredient to boost antioxidant activity, phenolic content, and shelf life of products such as bakery confectionery products, muffins, dairy products, natural colorants, beverage production, snack foods, as a natural preservative in meat products by reducing lipid oxidation, and for animal foods. These applications highlight the peel's versatility and value in enhancing food quality and sustainability.

8.1.13 Pomegranate seed powder

Both pomegranate seed and peel powders are rich in lysine, isoleucine, and sulfur-containing amino acids (methionine and cysteine), which are generally lacking in most foods, according to FAO/WHO reference protein standards. Seed pulp extract can be used as a natural dye in alcoholic beverages, adding flavor and aroma. In one study, supplementation with pomegranate seed powder (PSP) at 5 g twice daily was shown to significantly lower fasting blood glucose, glycated hemoglobin (HbA1c), total cholesterol, and triglycerides in patients with type 2 diabetes mellitus over an 8-week period compared to a placebo group. The results suggested PSP may benefit glucose and HbA1c levels, though its effects on triglycerides and cholesterol were less conclusive.

8.1.14 Pomegranate waste utilization

The pomegranate tree, including its roots, bark, leaves, flowers, rind, and seeds, holds medicinal, industrial, and cosmetic value. The rind, for example, is used in medicine, leather, and dye industries, as well as in tooth powder formulations. Advanced processing methods have achieved a 15.5% recovery of rind powder from the whole fruit and 34% from the rind weight. The bark also has medicinal applications, particularly in Ayurvedic formulations. Overall, pomegranate by-products are a valuable resource, rich in alkaloids, aromatic compounds, and enzymes, offering numerous industrial applications.

8.2 Conclusion

Overall, all fruits, such as *Ziziphus spina-christi* (Sidr), *Phoenix dactylifera* (Date palm), and *Punica granatum* (Pomegranate), are versatile fruits with numerous applications in functional foods. Their rich bioactive content, including phenolics, flavonoids, and antioxidants, makes them an ideal component for improving the nutritional profile of various food products. From yogurts to baked goods, and meat products to beverages, incorporating medicinal fruits and their byproducts enhances both the health-promoting and sensory properties of these foods, making them a significant functional ingredient in the food industry.

This book chapter on three medicinal fruits provides in-depth research and a detailed overview of their bioactive compounds, traditional medicinal uses, and potential applications in nutraceuticals and functional foods. The wealth of research highlights the importance of these fruits in traditional Kuwaiti medicine and modern food technology. Continued investigation into their processing methods and bioactive profiles will likely expand their use in health-promoting products. Advances in processing and product development have contributed to greater utilization and consumption of the fruit. Current research focuses on creating new, fruit-derived food products, including baked and minimally processed arils, juice, squash, beverages, molasses, concentrates, frozen seeds, jams, jellies, marmalades, grenadine, wine, and various powders and oils. However, the widespread adoption of these products has been hampered by a lack of commercially viable processing technologies. To address this, there is a pressing need to develop and standardize efficient methods, extraction processes, and storage solutions for products. Incorporating modern technologies could also enhance Kuwaiti medicinal fruit processing and product popularity. This chapter could build a scientific foundation for the successful commercialization and utilization of Kuwaiti medicinal fruit processing technologies.

References

Abdel-Maksoud, B. S., El-Waseif, M. A. E. M., Fahmy, H. M., Abd-Elazim, E. I., & Shaaban, H. A. G. (2022). Study effect of addition date seeds powder on quality criteria and antioxidant properties of beef meatballs. *Egyptian Journal of Chemistry*, *65*(11), 627–640.

Adhami, V. M., & Mukhtar, H. (2006). Polyphenols from green tea and pomegranate for prevention of prostate cancer. *Free Radical Research*, *40*(10), 1095–1104.

Ahmed, J., Al-Jasass, F. M., & Siddiq, M. (2013). Date fruit composition and nutrition. In: Siddiq, M., Aleid, A. M., & Kader, A. A. (Eds.), *Dates: Postharvest Science, Processing Technology and Health Benefits* (pp. 261–283). Chichester: Wiley Blackwell.

Akuru, E. A., Mpendulo, C. T., Oyeagu, C. E., & Nantapo, C. W. T. (2021). Pomegranate (Punica granatum L.) peel powder meal supplementation in broilers: effect on growth performance, digestibility, carcase and organ weights, serum and some meat antioxidant enzyme biomarkers. *Italian Journal of Animal Science*, *20*(1), 119–131.

Aleid, S. M., & Haddadin, J. S. (2023). Valorization and chemical constituents assessments of Khalas dates fruit, syrup and pits. *Current Research in Nutrition and Food Science Journal*, *11*(1), 77–87.

Al-Farsi, M., Alasalvar, C., Morris, A., Baron, M., & Shahidi, F. (2005). Comparison of antioxidant activity, anthocyanins, carotenoids, and phenolics of three native fresh and sun-dried date (Phoenix dactylifera L.) varieties grown in Oman. *Journal of Agricultural and Food Chemistry*, *53*(19), 7592–7599.

Al-Farsi, M. A., & Lee, C. Y. (2008). Nutritional and functional properties of dates: a review. *Critical Reviews in Food Science and Nutrition*, *48*(10), 877–887.

Almusallam, I. A., Mohamed Ahmed, I. A., Saleh, A., Al-Juhaimi, F. Y., Ghafoor, K., Al Maiman, S., & Babiker, E. E. (2021). Potential of date palm spikelet extract as an anti-oxidative agent in set-type yogurt during cold storage. *CyTA-Journal of Food*, *19*(1), 190–197.

Alper, N., Bahçeci, K. S., & Acar, J. (2005). Influence of processing and pasteurization on color values and total phenolic compounds of pomegranate juice. *Journal of Food Processing and Preservation*, *29*(5–6), 357–368.

Al-Turki, S., Shahba, M. A., & Stushnoff, C. (2010). Diversity of antioxidant properties and phenolic content of date palm (Phoenix dactylifera L.) fruits as affected by cultivar and location. *Journal of Food, Agriculture and Environment, 8*(1), 253–260.

Alyahya, S., Arif, M., Wasim, R., & Kushwaha, S. P. (2022). Phoenix dactylifera L. (Ajwa Dates): an energetic plant fruit with immense nutraceuticals value for strengthening the power of resistance. *Journal of Pharmacognosy and Phytochemistry, 11*(3), 151–162.

Atwaa, E. S. H., Shahein, M. R., Alrashdi, B. M., Hassan, M. A., Alblihed, M. A., Dahran, N., ... & Elmahallawy, E. K. (2022). Effects of fermented camel milk supplemented with Sidr fruit (Ziziphus spina-christi L.) pulp on hyperglycemia in streptozotocin-induced diabetic rats. *Fermentation, 8*(6), 269.

Aviram, M., Dornfeld, L., Rosenblat, M., Volkova, N., Kaplan, M., Coleman, R., ... & Fuhrman, B. (2000). Pomegranate juice consumption reduces oxidative stress, atherogenic modifications to LDL, and platelet aggregation: studies in humans and in atherosclerotic apolipoprotein E–deficient mice. *The American Journal of Clinical Nutrition, 71*(5), 1062–1076.

Aware, M. R., Perke, D. S., & Praneetha, Y. (2019). Trend and competitiveness of pomegranate in domestic and international prices. *Journal of Pharmacognosy and Phytochemistry, 8*(6), 2081–2083.

Ayad, A. A., Williams, L. L., Gad El-Rab, D. A., Ayivi, R., Colleran, H. L., Aljaloud, S., & Ibrahim, S. A. (2020). A review of the chemical composition, nutritional and health benefits of dates for their potential use in energy nutrition bars for athletes. *Cogent Food & Agriculture, 6*(1), 1809309.

Bekeir, H. S., Hamad, A., Eleiwa, N. Z., & Amin, R. A. (2024). Molecular Docking analysis of the antibacterial effect of pomegranate (Punica granatum) fruit molasses bioactive compounds. *Benha Veterinary Medical Journal, 46*(2), 90–95.

Benyahia-Krid, F. A., Khemissat, E., Zitoun, O. A., Djafri, K., Bergouia, M., Meghzili, B., & El-Mechta, L. (2021). Technological valorization of Algerian dates downgraded from deglet nour variety to semi-candied dates in dates syrup. *Food and Nutrition Sciences, 12*(6), 627–642.

Campos-Lozada, G., Hernández-Miranda, J., del Valle-Mondragón, L., Ortiz-Polo, A., Betanzos-Cabrera, G., & Aguirre-Álvarez, G. (2024). Effects of hyperbaric (non-thermal) sanitization and the method of extracting pomegranate juice on its antioxidant and antihypertensive properties. *Antioxidants, 13*(8), 1009.

Cano-Lamadrid, M., Calín-Sánchez, Á., Clemente-Villalba, J., Hernández, F., Carbonell-Barrachina, Á. A., Sendra, E., & Wojdyło, A. (2020). Quality parameters and consumer acceptance of jelly candies based on pomegranate juice "Mollar de Elche". *Foods, 9*(4), 516.

Chidambara Murthy, K. N., Reddy, V. K., Veigas, J. M., & Murthy, U. D. (2004). Study on wound healing activity of Punica granatum peel. *Journal of Medicinal Food, 7*(2), 256–259.

Dadashi, S., Mousazadeh, M., Emam-Djomeh, Z., & Mousavi, S. M. (2013). Pomegranate (Punica granatum L.) seed: a comparative study on biochemical composition and oil physicochemical characteristics. *International Journal of Advanced Biological and Biomedical Research, 1*(4), 351–363.

Dargham, M. B., Boumosleh, J. M., Farhat, A., Abdelkhalek, S., Bou-Maroun, E., & El Hosry, L. (2022). Antioxidant and anti-diabetic activities in commercial and homemade pomegranate molasses in Lebanon. *Food Bioscience, 46*, 101540.

Delgado, N. T., Rouver, W. N., & Dos Santos, R. L. (2020). Protective effects of pomegranate in endothelial dysfunction. *Current Pharmaceutical Design, 26*(30), 3684–3699.

Devhare, P. F., Kotecha, P. M., Godase, S. N., & Chavan, U. D. (2021). Studies on utilization of pomegranate juice in the preparation of peanut chikki. *International Journal of Chemical Studies, 9*(1), 1532–1535.

Djaoud, K., Boulekbache-Makhlouf, L., Yahia, M., Mansouri, H., Mansouri, N., Madani, K., & Romero, A. (2020). Dairy dessert processing: effect of sugar substitution by date syrup and powder on its quality characteristics. *Journal of Food Processing and Preservation, 44*(5), e14414.

Echegaray, N., Gullón, B., Pateiro, M., Amarowicz, R., Misihairabgwi, J. M., & Lorenzo, J. M. (2023). Date fruit and its by-products as promising source of bioactive components: a review. *Food Reviews International, 39*(3), 1411–1432.

Echegaray, N., Pateiro, M., Gullón, B., Amarowicz, R., Misihairabgwi, J. M., & Lorenzo, J. M. (2020). Phoenix dactylifera products in human health–a review. *Trends in Food Science & Technology, 105*, 238–250.

EH, A., Hanan, A. R., & Shalaby, H. S. (2021). Therapeutic potential of low-fat yogurt supplemented with Dried Nabaq fruit (Zizyphus spina-christi L) on hyperglycemia in streptozotocin-induced diabetic rats. *Egyptian Journal of Nutrition, 36*(2), 149–201.

Elanany, A. A., & Salem, A. (2024). Biological and technological evaluation of sidr (Ziziphus spina Christi L.) based high fat diet as antihyperlipidemic effects. *Food Technology Research Journal, 3*(2), 94–106.

Elsharawy, N. T., Al-Mutarrafi, M., & Al-Ayafi, A. (2019). Different types of dates in Saudi Arabia and its most fungal spoilage and its most preservation methods. *International Journal of Recent Scientific Research, 10*(11), 35787–35799.

Esposto, S., Veneziani, G., Taticchi, A., Urbani, S., Selvaggini, R., Sordini, B., … & Servili, M. (2021). Chemical composition, antioxidant activity, and sensory characterization of commercial pomegranate juices. *Antioxidants, 10*(9), 1381.

Famuwagun, A. A., & Gbadamosi, S. O. (2021). Nutritional properties, antioxidant and enzyme inhibitory activities of bread sweetened with date fruit. *Turkish Journal of Agriculture-Food Science and Technology, 9*(1), 114–123.

Gad, A. S., Kholif, A. M., & Sayed, A. F. (2010). Evaluation of the nutritional value of functional yogurt resulting from combination of date palm syrup and skim milk. *American Journal of Food Technology, 5*(4), 250–259.

Gaikwad, N. N., Kadam, A. A., Giri, N. A., Suryavanshi, S. K., & Marathe, R. A. (2024). Enhancing bioactive compounds retention in pomegranate juice powder through foam mat drying. *Journal of Food Process Engineering, 47*(8), e14707.

García-Pastor, M. E., Zapata, P. J., Castillo, S., Martínez-Romero, D., Guillén, F., Valero, D., & Serrano, M. (2020). The effects of salicylic acid and its derivatives on increasing pomegranate fruit quality and bioactive compounds at harvest and during storage. *Frontiers in Plant Science, 11*, 668.

Gil, M. I., Tomás-Barberán, F. A., Hess-Pierce, B., Holcroft, D. M., & Kader, A. A. (2000). Antioxidant activity of pomegranate juice and its relationship with phenolic composition and processing. *Journal of Agricultural and Food chemistry, 48*(10), 4581–4589.

Giri, N. A., Bhangale, A., Gaikwad, N. N., Manjunatha, N., Raigond, P., & Marathe, R. A. (2024). Comparative study on effect of pomegranate peel powder as natural preservative and chemical preservatives on quality and shelf life of muffins. *Scientific Reports*, *14*(1), 10307.

Habib, H. M., El-Gendi, H., El-Fakharany, E. M., El-Ziney, M. G., El-Yazbi, A. F., Al Meqbaali, F. T., & Ibrahim, W. H. (2023). Antioxidant, anti-inflammatory, antimicrobial, and anticancer activities of pomegranate juice concentrate. *Nutrients*, *15*(12), 2709.

Hassan, A., Ashfaq, J., Mudassar, M., Saeed, S., & Raza, M. Q. (2024). Evaluating the physiochemical and nutritional attributes of dates (Phoenix dactylifera) based cupcakes. *European Journal of Health Sciences*, *10*(1), 37–47.

Husain, M. I., Devikar, G. Y., Singh, G., & Jaiswal, P. K. (2004). Quality evaluation of pomegranate seed and powder. *Spice India*, *17*(7), 28–32.

Hussein, A. S. (2019). Ziziphus spina-christi: analysis of bioactivities and chemical composition. In: Mariod, A. A. (Ed.), Wild Fruits*: Composition, Nutritional Value and Products* (pp. 175–197). Cham: Springer.

Huxley, R. R., & Neil, H. A. W. (2003). The relation between dietary flavonol intake and coronary heart disease mortality: a meta-analysis of prospective cohort studies. *European Journal of Clinical Nutrition*, *57*(8), 904–908.

Ilce Gabriela, M. M., & Girish, G. (2017). Fruit processing by-products: a rich source for bioactive compounds and value added products. In: Anal, A. (Ed.), *Food Processing By-products and Their Utilization* (pp. 11–26). New Jersey: John Wiley & Sons, Inc.

Jalal, H., Pal, M. A., Hamdani, H., Rovida, M., & Khan, N. N. (2018). Antioxidant activity of pomegranate peel and seed powder extracts. *Journal of Pharmacognosy and Phytochemistry*, *7*(5), 992–997.

Jurenka, J. (2008). Therapeutic applications of pomegranate (Punica granatum L.): a review. *Alternative Medicine Review*, *13*(2), 128–144.

Kamal-Eldin, A., Hashim, B. I., & Mohamed, O. I. (2012). Processing and utilization of palm date fruits for edible applications. *Recent Patents on Food, Nutrition & Agriculture*, *4*(1), 78–86.

Karizaki, V. M. (2017). Iranian dates and ethnic date-based products. *Journal of Ethnic Foods*, *4*(3), 204–209.

Kokkinomagoulos, E., Nikolaou, A., Kourkoutas, Y., & Kandylis, P. (2020). Evaluation of yeast strains for pomegranate alcoholic beverage production: effect on physicochemical characteristics, antioxidant activity, and aroma compounds. *Microorganisms*, *8*(10), 1583.

Kola, O., Parildi, E., Akkaya, M., Özcan, B. D., Mamur, E., & Turker, B. (2021). Fatty acids, sterols and triglycerides composition of cold pressed oil from pomegranate seeds. *Rivista Italiana delle Sostanze Grasse*, *98*(3), 197–204.

Kushwaha, S. C., Bera, M. B., & Kumar, P. (2013). Nutritional composition of detanninated and fresh pomegranate peel powder. *IOSR Journal of Environmental Science, Toxicology and Food Technology*, *7*(1), 38–42.

Lacivita, V., Incoronato, A. L., Conte, A., & Del Nobile, M. A. (2021). Pomegranate peel powder as a food preservative in fruit salad: a sustainable approach. *Foods*, *10*(6), 1359.

Lashkari, H., Varidi, M. J., Eskandari, M. H., & Varidi, M. (2020). Effect of pomegranate juice on the manufacturing process and characterization of feta-type cheese during storage. *Journal of Food Quality*, *2020*(1), 8816762.

Liu, G., Xu, X., Hao, Q., & Gao, Y. (2009). Supercritical CO_2 extraction optimization of pomegranate (Punica granatum L.) seed oil using response surface methodology. *LWT-Food Science and Technology*, *42*(9), 1491–1495.

López-Rubira, V., Conesa, A., Allende, A., & Artés, F. (2005). Shelf life and overall quality of minimally processed pomegranate arils modified atmosphere packaged and treated with UV-C. *Postharvest Biology and Technology*, *37*(2), 174–185.

Maestre, J., Melgarejo, P., Tomas-Barberan, F. A., & Garcia-Viguera, C. (2000). New food products derived from pomegranate. In: Moreno, P. M. (Ed.), *Production, Processing and Marketing of Pomegranate in the Mediterranean Region: Advances in Research and Technology* (pp. 243–245). Paris: Centre international de hautes études agronomiques méditerranéennes.

Malik, A., Afaq, F., Sarfaraz, S., Adhami, V. M., Syed, D. N., & Mukhtar, H. (2005). Pomegranate fruit juice for chemoprevention and chemotherapy of prostate cancer. *Proceedings of the National Academy of Sciences*, *102*(41), 14813–14818.

Manickavasagan, A., Mathew, T. A., Al-Attabi, Z. H., & Al-Zakwani, I. M. (2013). Dates as a substitute for added sugar in traditional foods-a case study with idli. *Emirates Journal of Food & Agriculture (EJFA)*, *25*(11), 899–906.

Mantzourani, I., Terpou, A., Bekatorou, A., Mallouchos, A., Alexopoulos, A., Kimbaris, A., … & Plessas, S. (2020). Functional pomegranate beverage production by fermentation with a novel synbiotic L. paracasei biocatalyst. *Food Chemistry*, *308*, 125658.

Maskan, M. (2006). Production of pomegranate (Punica granatum L.) juice concentrate by various heating methods: colour degradation and kinetics. *Journal of Food Engineering*, *72*(3), 218–224.

Meghwar, P., Shar, N. A., Shaikh, S. A., Baloch, A. B., Kumar, D., Dhanker, P., & Depar, A. A. (2021). Traditional process of dates, COVID-19 pandemic observation, and challenges in Sindh: dates traditional processing techniques, and challenges. *Proceedings of the Pakistan Academy of Sciences: B. Life and Environmental Sciences*, *58*(3), 83–94.

Mena, P., Gironés-Vilaplana, A., Martí, N., & García-Viguera, C. (2012). Pomegranate varietal wines: phytochemical composition and quality parameters. *Food Chemistry*, *133*(1), 108–115.

Mena, P., Vegara, S., Martí, N., García-Viguera, C., Saura, D., & Valero, M. (2013). Changes on indigenous microbiota, colour, bioactive compounds and antioxidant activity of pasteurised pomegranate juice. *Food Chemistry*, *141*(3), 2122–2129.

Mirza, M. B., Elkady, A. I., Al-Attar, A. M., Syed, F. Q., Mohammed, F. A., & Hakeem, K. R. (2018). Induction of apoptosis and cell cycle arrest by ethyl acetate fraction of Phoenix dactylifera L. (Ajwa dates) in prostate cancer cells. *Journal of Ethnopharmacology*, *218*, 35–44.

Mohapatra, S., Bhanj, P. U., Roy, M., & Das, A. (2023). Development of nutritionally enhanced pomegranate peel powder by blending with certain spices. *Environment and Ecology*, *41*(2), 802–806.

Mousavi, T., Rafiei, A., & Yoosefpour, M. (2014). Nutritional value and health benefits of dates according to Islamic recourses and traditional medicine. *Journal of Mazandaran University of Medical Sciences*, *24*(117), 247–265.

Mrabet, A., Rodríguez-Gutiérrez, G., Guillén-Bejarano, R., Rodríguez-Arcos, R., Ferchichi, A., Sindic, M., & Jiménez-Araujo, A. (2015). Valorization of Tunisian secondary date varieties (Phoenix dactylifera L.) by hydrothermal treatments: new fiber concentrates with antioxidant properties. *LWT-Food Science and Technology*, *60*(1), 518–524.

Nasir, M. U., Hussain, S., Jabbar, S., Rashid, F., Khalid, N., & Mehmood, A. (2015). A review on the nutritional content, functional properties and medicinal potential of dates. *Scientific Letters*, *3*(1), 17–22.

Nuncio-Jáuregui, N., Calín-Sánchez, Á., Vázquez-Araújo, L., Pérez-López, A. J., Frutos-Fernández, M. J., & Carbonell-Barrachina, Á. A. (2015). Processing pomegranates for juice and impact on bioactive components. In: Preedy, V. R. (Ed.), *Processing and Impact on Active Components in Food* (pp. 629–636). London, UK: Academic Press.

Nwanekezi, E. C., Ekwe, C. C., & Agbugba, R. U. (2015). Effect of substitution of sucrose with date palm (Phoenix dactylifera) fruit on quality of bread. *Journal of Food Processing & Technology*, *6*(9), 1000484.

Omer, H. A., Abdel-Magid, S. S., & Awadalla, I. M. (2019). Nutritional and chemical evaluation of dried pomegranate (Punica granatum L.) peels and studying the impact of level of inclusion in ration formulation on productive performance of growing Ossimi lambs. *Bulletin of the National Research Centre*, *43*, 1–10.

Orak, H. H. (2009). Evaluation of antioxidant activity, colour and some nutritional characteristics of pomegranate (Punica granatum L.) juice and its sour concentrate processed by conventional evaporation. *International Journal of Food Sciences and Nutrition*, *60*(1), 1–11.

Palabiyik, I., Toker, O. S., Konar, N., Gunes, R., Güleri, T., Alaşalvar, H., & Çam, M. (2018). Phenolics release kinetics in sugared and sugar-free chewing gums: microencapsulated pomegranate peel extract usage. *International Journal of Food Science & Technology*, *53*(12), 2657–2663.

Palak, T., Ranu, P., & Anisha, V. (2020). Sensory analysis of pomegranate peel powder in the development of value added food products. *International Journal of Scientific Research*, *9*, 414–417.

Pérez-Vicente, A., Serrano, P., Abellán, P., & García-Viguera, C. (2004). Influence of packaging material on pomegranate juice colour and bioactive compounds, during storage. *Journal of the Science of Food and Agriculture*, *84*(7), 639–644.

Qin, Y. Y., Zhang, Z. H., Li, L., Xiong, W., Shi, J. Y., Zhao, T. R., & Fan, J. (2013). Antioxidant effect of pomegranate rind powder extract, pomegranate juice, and pomegranate seed powder extract as antioxidants in raw ground pork meat. *Food Science and Biotechnology*, *22*, 1063–1069.

Ranjitha, J., Bhuvaneshwari, G., Terdal, D., & Kavya, K. (2018). Nutritional composition of fresh pomegranate peel powder. *International Journal of Chemical Studies*, *6*(4), 692–696.

Rios-Corripio, G., Welti-Chanes, J., Rodriguez-Martinez, V., & Guerrero-Beltrán, J. Á. (2020). Influence of high hydrostatic pressure processing on physicochemical characteristics of a fermented pomegranate (Punica granatum L.) beverage. *Innovative Food Science & Emerging Technologies*, *59*, 102249.

Rowayshed, G., Salama, A., Abul-Fadl, M., Akila-Hamza, S., & Emad, A. M. (2013). Nutritional and chemical evaluation for pomegranate (Punica granatum L.) fruit peel and seeds powders by products. *Middle East Journal of Applied Sciences, 3*(4), 169–179.

Sadeghi, N., Janat, B., Oveysi, M., Haji, M. M., & Fotovat, M. (2009). Antioxidant activity of Iranian pomegranate (Punica granatum L.) seed extracts. *Journal of Agricultural Science and Technology, 11*(5), 633–638.

Salomón-Torres, R., Valdez-Salas, B., & Norzagaray-Plasencia, S. (2021). Date palm: source of foods, sweets and beverages. In: Al-Khayri, J. M., Jain, S. M., Johnson, D. V. (Eds.), *The Date Palm Genome, Vol. 2: Omics and Molecular Breeding* (pp. 3–26). Cham: Springer International Publishing.

Schubert, S. Y., Lansky, E. P., & Neeman, I. (1999). Antioxidant and eicosanoid enzyme inhibition properties of pomegranate seed oil and fermented juice flavonoids. *Journal of Ethnopharmacology, 66*(1), 11–17.

Shahamirian, M., Eskandari, M. H., Niakousari, M., Esteghlal, S., Hashemi Gahruie, H., & Mousavi Khaneghah, A. (2019). Incorporation of pomegranate rind powder extract and pomegranate juice into frozen burgers: oxidative stability, sensorial and microbiological characteristics. *Journal of Food Science and Technology, 56*, 1174–1183.

Shahein, M. R., Atwaa, E. S. H., Elkot, W. F., Hijazy, H. H. A., Kassab, R. B., Alblihed, M. A., & Elmahallawy, E. K. (2022). The impact of date syrup on the physicochemical, microbiological, and sensory properties, and antioxidant activity of bio-fermented camel milk. *Fermentation, 8*(5), 192.

Siddiq, M., & Greiby, I. (2013). Overview of date fruit production, postharvest handling, processing, and nutrition. In: Siddiq, M., Aleid, A. M., & Kader, A. A. (Eds.), *Dates: Postharvest Science, Processing Technology and Health Benefits* (pp. 1–28). Chichester: Wiley Blackwell.

Singh, R. P., Chidambara Murthy, K. N., & Jayaprakasha, G. K. (2002). Studies on the antioxidant activity of pomegranate (Punica granatum) peel and seed extracts using in vitro models. *Journal of Agricultural and Food Chemistry, 50*(1), 81–86.

Stover, E. D., & Mercure, E. W. (2007). The pomegranate: a new look at the fruit of paradise. *HortScience, 42*(5), 1088–1092.

Sumner, M. D., Elliott-Eller, M., Weidner, G., Daubenmier, J. J., Chew, M. H., Marlin, R., ... & Ornish, D. (2005). Effects of pomegranate juice consumption on myocardial perfusion in patients with coronary heart disease. *The American Journal of Cardiology, 96*(6), 810–814.

Tawfek, M. A., Hussein, M. T., & Rizk, A. E. (2022). Low-fat ice milk manufactured with fruits of NABQ (Ziziphus spina-christi L.). *Carpathian Journal of Food Science and Technology*. https://doi.org/10.34302/crpjfst/2022.14.3.4

Thakur, N. S., Dhaygude, S., & Joshi, V. K. (2013). Development of wild pomegranate aril-in-syrup and its quality evaluation during storage. *International Journal of Food and Fermentation Technology, 3*(1), 33–40.

Velotto, S., Palmeri, R., Alfeo, V., Gugino, I. M., Fallico, B., Spagna, G., & Todaro, A. (2023). The effect of different technologies in pomegranate jam preparation on the phenolic compounds, vitamin C and antioxidant activity. *Food Bioscience, 53*, 102525.

Vijayanand, P., & Kulkarni, S. G. (2012). Processing of dates into value-added products. In: Mohamed Essa, M., Manickavasagan, A., & Sukumar, E. (Ed.), *Dates Production, Processing, Food and Medicinal Values* (pp. 255–264). Boca Raton, FL: CRC Press.

Wang, R. F., Xie, W. D., Zhang, Xing, D. M., Ding, Y., Wang, W., … & Du, L. J. (2004). Bioactive Compounds from the Seeds of Punica g ranatum (Pomegranate). *Journal of Natural Products, 67*(12), 2096–2098.

Yadav, K., Sarkar, B. C., & Kumar, P. (2006). Pomegranate: recent developments in harvesting, processing and utilization. *Indian Food Industry, 25*(2), 56–63.

Yun, Y., Li, J., Pan, F., Zhou, Y., Feng, X., Tian, J., … & Zhou, L. (2023). A novel strategy for producing low-sugar pomegranate jam with better anthocyanin stability: combination of high-pressure processing and low methoxyl & amidated pectin. *LWT, 179*, 114625.

chapter nine

Bioavailability of therapeutic constituents of medicinal fruits of Jordan

Meenakshisundaram Selvamuthukumaran

9.1 Introduction

Medicinal fruits existing in Jordan contain a wide array of bio-active constituents; the presence of such bio-active constituents leads to exhibiting several therapeutic benefits. The therapeutic benefits can be exerted if the bio-active constituents are readily available, i.e., if they are absorbed and metabolized, and after the metabolization process, such constituents should be available to exert significant benefits to the consumers. The availability of such therapeutic components derived from consuming such medicinal fruits is explained below.

9.2 Bioavailability of therapeutic constituents of medicinal fruits available in Jordan

9.2.1 Pomegranate

The fruit *Punica granatum* contains several polyphenols, which have excellent therapeutic activity. The functional properties of polyphenolic components can be achieved if they are absorbed efficiently and should be readily available in the internal organs, either in the intestine, kidney, liver, heart, or even in tissues, i.e., in blood, to exert such therapeutic effect.

Seeram et al. (2004) conducted in vivo studies to demonstrate, especially, the role of pomegranate fruit antioxidants like ellagic acid and hydrolyzable ellagitannins in exhibiting biological properties based on bioavailability. They chose a human model for their study and fed the pomegranate juice at 180 mL, which contained antioxidants like ellagic acid at 25 mg and hydrolyzable ellagitannins at 318 mg. As a result of in vivo feeding, they observed that ellagic acid was detected in the human plasma at a concentration of 31.9 ng/mL after 1 hour of ingestion of pomegranate

 DOI: 10.1201/9781003452614-9

juice, but it remained bioavailable for up to a period of 4 hours. They further concluded that medicinal properties can be achieved after consuming pomegranate juice, ascribed to its bioavailability for a specific time period. In another study, pomegranate juice was ingested by diabetic patients at 250 mL/day, which is equivalent to 345.87 μg/mL of catechin equivalents. Their study confirms that the ingestion of pomegranate juice significantly enhanced the total antioxidant capacity and, at the same time, reduced the concentration of MDA (Malondialdehyde) (Sohrab et al., 2015). The ingestion of pomegranate juice led to the storage of polyphenolic compounds, especially catechin, in the system, which led to the exhibition of bio-active properties, thereby significantly enhancing the antioxidant capacity. Pomegranate juice was administered to both humans and rodents for a time period of 11 weeks. The administered dose of 6.25 mL of pomegranate juice contained a polyphenol concentration of 0.175 mmol, while 12.5 mL of juice contained a 0.35 mmol concentration of total polyphenols. As a result of feeding, the authors observed that the pomegranate juice minimized the vessel damage caused by atherosclerosis by 44% (Aviram et al., 2000).

Sohrab et al. (2017) reported that the consumption of pomegranate juice at 200 mL/day, which is equivalent to total polyphenols of 2.12 g and flavonoids of 0.03 g for a period of 6 weeks, resulted in a reduction of oxidized LDL and enhanced the serum level of total antioxidant capacity among the individuals. This demonstrates that the availability of flavonoids and polyphenols derived from the administration of pomegranate juice resulted in a reduction of LDL levels in the blood. Similarly, HDL levels were enhanced as a result of the consumption of pomegranate juice, around 50 mL containing polyphenol concentration at 6.3 mg/100 g for a period of 1 month (Shishehbor et al., 2016) (Table 9.1).

9.2.2 *Fig*

Fig also contains several bio-active constituents, which can contribute to excellent medicinal value. The fruit skin and seed also contain a significant quantity of polyphenolic constituents, which have a lot of functional value. The wide array of polyphenols can be quantified from fig fruit, as shown in Figure 9.1.

9.2.2.1 *Bioavailability of fig polyphenols by in vitro gastrointestinal model system*

The various research works describe that the evaluation of bio-active constituents using in vitro models can be correlated well with in vivo models by either using humans or animals as test subjects (Bouayed et al., 2011).

Kaminoglu and Capanoglu (2014) evaluated the bio-availability of various phenolic compounds, which include total phenols, flavonoids, and

Table 9.1 Bioaccessibility of phenolic compounds from pomegranate

Name of the polyphenolic compound	Bioavailability status	Functional effect achieved	References
Flavonoids and phenols	Blood	Reduction of LDL levels in blood	Sohrab et al. (2017)
Ellagic acid and hydrolyzable ellagitannins	Blood	Antioxidant activity	Seeram et al. (2004)
Catechin	Blood	Enhanced the total antioxidant capacity and reduced the concentration of MDA	Sohrab et al. (2015)
Total phenols	Blood	Reduced LDL oxidation up to 90% and the vessel damage caused by atherosclerosis to 44%	Aviram et al. (2000)

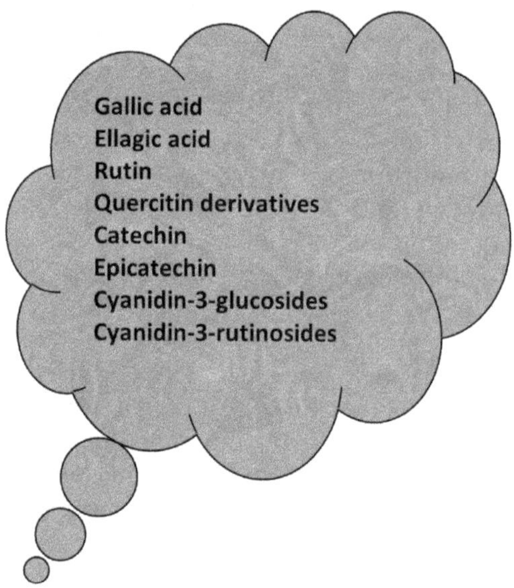

Figure 9.1 Polyphenolic constituents from fig fruit.

anthocyanins, in two different fig cultivars using an in vitro gastrointestinal model system. Their findings reveal that they were able to detect total phenols of around 22% and 32% after digesting with pepsin, while total antioxidant capacity increased to 80% and 86%, and total flavonoids increased to 17% and 32% for yellow and purple-colored fig fruit cultivars, exhibiting the availability of functional components in order to exert therapeutic benefits for the consumers.

9.3 Conclusion

In order to exert therapeutic benefits, it is essential to retain the various bio-active constituents in the internal organs so that the desired functional effect can be achieved successfully.

References

Aviram, M., Dornfeld, L., Rosenblat, M., Volkova, N., Kaplan, M., Coleman, R., Hayek, T., Presser, D., & Fuhrman, B. Pomegranate juice consumption reduces oxidative stress, atherogenic modifications to LDL, and platelet aggregation: Studies in humans and in atherosclerotic apolipoprotein E–deficient mice. *Am. J. Clin. Nutr.* 2000, *71*, 1062–1076.

Bouayed, J., Hoffman, L., & Bohn, T. Total phenolics, flavonoids, anthocyanins, and antioxidant activity following simulated gastro-intestinal digestion and dialysis of apple varieties: Bioaccessibility and potential uptake. *Food Chem.* 2011, *128*, 14–21.

Kamiloglu, S., & Capanoglu, E. Polyphenol content in figs (*Ficus carica* L.): Effect of sun-drying. *Int. J. Food Prop.* 2014, *18*(3), 521–535.

Seeram, N., Lee, R., & Heber, D. Bioavailability of ellagic acid in pomegranate (Punica granatum L.) juice. *Clin. Chim. Acta Int. J. Clin. Chem.* 2004, *348*, 63–68. doi:10.1016/j.cccn.2004.04.029

Shishehbor, F., Zarei, M., Saki, A., Zakerkish, M., Shirani, F., & Zare, M. Effects of concentrated pomegranate juice on subclinical inflammation and cardiometabolic risk factors for type 2 diabetes: A quasi-experimental study. *Int. J. Endocrinol. Metab.* 2016, *14*, e33835.

Sohrab, G., Angoorani, P., Tohidi, M., Tabibi, H., Kimiagar, M., & Nasrollahzadeh, J. Pomegranate (Punicagranatum) juice decreases lipid peroxidation, but has no effect on plasma advanced glycated end-products in adults with type 2 diabetes: A randomized double-blind clinical trial. *Food Nutr. Res.* 2015, *59*, 28551.

Sohrab, G., Ebrahimof, S., Sotoudeh, G., Neyestani, T. R., Angoorani, P., Hedayati, M., & Siasi, F. Effects of pomegranate juice consumption on oxidative stress in patients with type 2 diabetes: A single-blind, randomized clinical trial. *Int. J. Food Sci. Nutr.* 2017, *68*, 249–255.

chapter ten

Traditional/ethnic food preparations from medicinal fruits of Egypt

Arjun Mohanakumar, Rohini Dhenge, and Massimiliano Rinaldi

10.1 Introduction

Egypt, a nation strategically located in the northeastern corner of Africa, is bordered by the Mediterranean Sea to the north and the Red Sea to the east. This geographic positioning has not only shaped its climate and resources but has also made it a cultural crossroads between Africa, Asia, and Europe. Egypt's millennia-old history is a tapestry woven from the legacies of the Pharaohs, the monumental Pyramids of Giza, and the rich traditions that have evolved along the fertile banks of the Nile River. This ancient land, known as the cradle of civilization, has seen the rise and fall of empires, each leaving its mark on the country's diverse cultural and culinary heritage. Today, Egypt's historical richness continues to influence its people, particularly in the realm of traditional food preparations, where the use of medicinal fruits reflects the enduring connection between past and present. For millennia, Egypt's strategic location and the fertile lands of the Nile River Delta have sustained a thriving agricultural system, enabling the cultivation of grain crops, lush green vegetation, and the raising of livestock and poultry for meat, milk, wool, and other essential food products. These resources were crucial to the emergence and flourishing of Egyptian civilization, playing a pivotal role in shaping its Pharaonic history and the development of a rich and diverse food culture. Despite enduring successive waves of colonization by various foreign powers, Egypt has largely retained its distinct ethnic, linguistic, and cultural identities, preserving a meticulously documented and deeply rooted culinary heritage. This enduring legacy of traditional food culture, enriched by centuries of history, continues to be a vital aspect of Egypt's national identity today.

DOI: 10.1201/9781003452614-10

Ancient Egyptian gastronomy encompassed various aspects, including ingredient selection, kitchen organization, cooking tools, banquet preparation, and the rituals surrounding food consumption. Historical records, such as tomb depictions and hieroglyphic texts, reveal that ancient Egyptians were meticulous about their diet for two main reasons. First, they believed that proper nourishment was essential for achieving a long life. Second, they considered their earthly diet crucial for preparing for the afterlife, which they envisioned as a mirror of their life on earth (Mehdawy and Hussein, 2010). Studies were conducted by researchers on the carbon data of mummies that had lived in Egypt between 3500 BC and 600 AD to uncover what type of food they consumed (Laborde, 2017); The study indicated that ancient Egyptian diets varied by age group, with isotopic analysis revealing distinct chemical differences between the diets of children and adults. These differences were linked to the consumption of milk or millet porridge during infancy and childhood. The analysis also identified the most consumed foods, which were largely plant-based, including eggplants, pears, lentils, garlic, and wheat. These findings suggest that the ancient Egyptian diet was predominantly vegetarian.

Egyptian identity is shaped by two key elements. Firstly, Egypt's geographical location uniquely positions it as part of Europe, Asia, and Africa. Secondly, its rich history encompasses Pharaonic, Coptic, Greco-Roman, and Arabic Muslim influences. This blend of cultural origins significantly impacts Egyptian cuisine, making it diverse and vibrant. In northern Egypt, the cuisine reflects a mix of Greek and Roman influences, alongside those from foreign migrations, the Islamic conquest, and the Ottoman period. These historical and cultural interactions have enriched the culinary heritage and led to various recipe adaptations in the region. In contrast, southern Egypt (Upper Egypt and Nubia) has seen minimal external influence, preserving culinary traditions and ingredients that have remained largely unchanged since Pharaonic times due to the area's isolation from foreign cultures (Mehdawy and Hussein, 2010).

The Israelites recall six types of food they consumed in Egypt in the following sequence: daga, kishuim, avatiḥim, ḥatzir, betzalim, and shumim. These foods are initially categorized based on their natural origins—with daga (fish) representing the animal kingdom and the remaining items derived from plants. The plant-based foods can be further divided into two distinct categories: field crops from the Cucurbitaceae family (kishuim and avatiḥim) and pungent vegetables (ḥatzir, betzalim, and shumim) (Shemesh et al., 2020). Food habits, encompassing practices such as procurement, preparation, cooking, preservation, and consumption, are often passed down through generations as sacred traditions and family customs. Consequently, ethnic groups take great pride in preserving these culinary practices, which serve as a unifying force within communities (Parsa, 1998). As food heritage is increasingly recognized as a valuable legacy, there is growing concern and focused efforts on conserving traditional cuisines, which are deeply intertwined with the ingredients, production methods, dishes, and eating habits of a culture. However, preserving this cultural heritage presents challenges, as it is closely linked to the values, beliefs, behaviors, and societal norms that are deeply ingrained within a community (Ramli et al., 2016).

10.2 Egyptian medicinal fruits

A significant portion of the population in developing countries relies on plant-based resources for healthcare. While allopathic medicine is effective in treating a broad spectrum of diseases, its high cost and potential side effects are prompting many people to turn back to herbal medicines, which are often perceived to have fewer side effects (Mahmoud and Gairola, 2013). Medicinal plants play a vital role in the ethnomedicine of Egypt, with over 342 species being harvested from the wild. However, research indicates a concerning decline in these valuable resources (Elsabawy, 2012). Table 10.1 represents the species and families of medicinal fruits used for different uses or treatments (Mahmoud and Gairola, 2013).

Over the past three decades, the global market for herbs and herbal products has experienced significant growth. According to the Secretariat of the Convention on Biological Diversity, global sales of herbal products reached approximately US$60 billion in 2002. Moreover, Egypt boasts rich agricultural traditions and extensive biodiversity. The export of Egyptian medicinal and aromatic plants has become a significant industry and an important source of income worldwide (Abdel-Azim et al., 2011). While Egypt may not currently be among the leading exporters of medicinal plants, it remains a notable player in the global market for medicinal and aromatic plants. With an annual trade volume exceeding $60 billion, Egypt's exports amount to approximately $158.94 million, highlighting its significant role in this industry (El-Hamid et al., 2019). In a study conducted by Amer and Mohammad (2022) it was stated that medicinal plants are those that contain compounds valuable for therapeutic applications or for the synthesis of pharmaceuticals. Plants traditionally used in folk medicine, though lacking scientific validation, are considered crude drugs of natural or organic origin. These plants can be further categorized as "organized drugs" if they are derived from specific plant parts like roots, leaves, and stems, and as "unorganized drugs" if they are extracted from acellular parts of plants like oil, gums, gel, balsam, and exudates. Table 10.2 is provided above, which represents the traditional uses of the medicinal plant parts.

Table 10.1 Medicinal fruits in Egypt

Species	Family	Growth habit	Used part	Use(s)/ailment treated
Balanites aegyptiaca Del.	Zygophyllaceae	Tree	Fruits	Antidiabetic, antihelminthic
Capparis spinosa L.	Capparaceae	Shrub	Fruits	Treatment for catarrh
Citrullus colocynthis (L.) Schrad	Cucurbitaceae	Herb	Leaves, root, fruit	Diuretic, Rheumatism, purgative, carminative, antiepileptic
Ficus palmata Forssk.	Moraceae	Tree	Fruit	Demulcent, emollient, laxative, and poultice
Hyphaene thebaica (L.) Mart	Arecaceae	Tree	Fruit	Used in making soft drinks, molasses
Nitraria retusa (Forssk.)	Nitrariaceae	Shrub	Fruits	Urinary tract ailments
Rhus tripartita (Ucria)	Anacardiaceae	Tree	Fruits, leaves	Treatment for gastric and intestinal ailments
Salvadora persica L.	Salvadoraceae	Tree	Roots, branches, leaves, fruit	Used as a toothbrush and mouth antiseptic, also treats urinary tract pain and acts as a diuretic
Tribulus terrestris L.	Zygophyllaceae	Herb	Fruits	Natural nutritional supplement, helps with impotence
Ziziphus spina-christi (L.) Desf.	Rhamnaceae	Tree	Leaves	Treats colds, hypertension, and has astringent, anthelmintic, antidiarrheal, and demulcent properties
Zygophyllum coccineum L.	Zygophyllaceae	Herb	Fruits, seeds	Anthelmintic, diuretic
Zygophyllum simplex L.	Zygophyllaceae	Herb	Leaves, seeds	Anthelmintic

Table 10.2 Name of various plant parts that are grown in Egypt

Plant part	Description	Examples
Roots	Solid, fibrous, or fleshy underground parts of plants	Ginseng, stinging nettle, devil's claw roots
Rhizome	Fleshy stems growing horizontally underground, with leaves growing above ground	Ginger, spear grass, turmeric
Tubers	Swollen, fleshy underground structures	Potatoes, yams
Bark	Outer layers of trunks or stems, providing protection	Quinine bark, oak bark, pepper bark, willow bark
Leaves, stems, twigs, and buds	Various plant parts including leaves, stems, twigs, and buds used in herbal medicine	Maidenhair tree
Flower parts	Parts of flowers used for their medicinal properties	Pyrethrum, saffron, santonica, clove, chamomile
Seeds	Seeds of plants known for their medicinal properties	Bitter almond, black mustard, cardamom, ispaghula, kaladana, linseed, nutmeg
Fruits	Fruits of plants with medicinal properties	Amla, anise, bael, bahera, bitter orange, capsicum, caraway, cardamom, colocynth, coriander, cumin, dill, fennel, star anise, tamarind

10.2.1 *Baobab (*Adansonia digitata*)*

Baobab trees hold substantial aesthetic and ecological value, and they also contribute significantly to the livelihoods of rural communities. Many individuals in developing countries rely on non-timber forest products (NTFPs) for their economic well-being. NTFPs play a crucial role in rural areas by offering essential resources such as medicine, food, shelter, and providing spiritual and cultural benefits. Lisao et al. (2017) state that the traditional knowledge about the baobab is transmitted within families from older to younger generations; however, sharing this knowledge with individuals outside the family often requires a foundation of trust. Regardless of age, educational background, or economic status, all respondents acknowledged the baobab as a plant of significant value with important traditional uses. This widespread recognition underscores the plant's importance in regions where it is found, particularly as a vital food source that enhances the diet of rural communities. The pulp of ripe fruits is separated from the seeds and fibers by kneading it in cold water, creating an emulsion that is strained through a sieve. This emulsion is then mixed with thicker grain-based dishes to produce thinner porridge-like meals. The dried pulp can be eaten fresh or added to the porridge after cooking, which helps preserve its vitamin content. Additionally, the pulp is often ground to create a refreshing beverage with a flavor reminiscent of wine gums (Zahra'u et al., 2014). Baobab pulp is undeniably a rich source of vitamin C. By enhancing its handling and storage stability through appropriate techniques, its value could be significantly increased, potentially boosting interest in its use and contributing to improved food chain organization in developing countries such as Ethiopia, Nigeria, Tanzania, Kenya, Egypt, Ghana, and India. Currently, the preservation of baobab pulp remains largely uncontrolled, resulting in significant losses. Addressing the challenges of extending its shelf-life is crucial to maintaining its nutritional benefits and sensory qualities. Additionally, bioavailability studies are essential to better understand its health benefits, as not all dietary intake is fully absorbed by the human body (Chadare, 2010).

10.2.2 *Wine grapes (*Vitis vinifera*)*

The Leipzig University Museum houses the most extensive and well-preserved ancient papyrus focused on healing practices, known as the Ebers Papyrus. It was first studied and owned by Georg Ebers, who conducted scientific research on it. This medical papyrus consists of over 33 individual "books" or scrolls that were compiled into a single volume. Measuring 20.23 m in length, the papyrus dates back to the time of Pharaoh Amenhotep, around 1515 BC, according to the *Oxford History of Ancient Egypt*. The treatment for liver ailments in ancient Egypt involved a mixture of various natural ingredients: 1/8 of a fig, 1/8 of a desert plum, 1/16 of grapes, 1/8 of a sliced sycamore fig, 1/16 of ground pumpkin, 1/32 of acacia gum, 1/64 of incense, and 1/64 of cress. This mixture was consumed for four consecutive days. Such remedies were commonly used to address liver conditions, which were widespread due to the regular consumption of wine and fermented beverages like beer and fruit juices in place of water. The scarcity of fresh drinking water, the flooding of the Nile, and the prevalence of parasitic infections like schistosomiasis (*Schistosoma haematobium, Schistosoma mansoni*) further exacerbated cases of liver intoxication. Additionally, feasts of excess, particularly those dedicated to the goddess Hathor, often contributed to the frequency of such ailments. Several significant studies have demonstrated that this valuable plant offers positive effects on various aspects of metabolic syndrome, including dyslipidemia, diabetes, hypertension, and obesity. Additionally, it has shown cardioprotective and hepatoprotective properties in various experimental models. Given that grapes can be cultivated in diverse regions and are available in multiple forms, their numerous beneficial properties make them a promising option for the treatment of metabolic syndrome. The liver plays a crucial role in the metabolism of grape polyphenols, as these metabolites are present in high concentrations within the organ. Compounds like catechin and epicatechin are transformed in the liver into their methylated, sulfated, and glucuronidated forms. Non-metabolized catechin and epicatechin tend to accumulate in mesenteric white adipose tissue, while conjugated compounds are found in this tissue at lower concentrations. Limited research exists on the presence of flavanols and their metabolites in adipose tissue, although the free form of dimeric procyanidin *B2* has previously been detected there (Serra et al., 2011). Various clinical approaches exist for treating metabolic disorders, but among them, the use of functional foods stands out as a key therapeutic strategy. These foods, integrated seamlessly into daily diets, are easily consumed by patients with minimal risk of adverse effects. Traditional medicinal fruits like grapes, rich in natural polyphenols, are particularly noteworthy in this context. Whole grains, legumes, fruits, vegetables, and nuts are recognized as vital functional foods, offering significant health benefits. The established role of polyphenols in managing metabolic disorders further highlights the value of these traditional foods. Grapes, with their high polyphenol content, emerge as a promising natural remedy for addressing metabolic syndrome, blending the benefits of traditional dietary practices with modern health needs.

10.2.3 *Dates (*Phoenix dactylifera*)*

The growth of date fruits is impacted by a range of external and internal factors, which can be primarily categorized based on different physical and chemical changes, such as alterations in color at various stages of development (Khalid et al., 2017). Dates can be eaten at different maturity stages: Khalal (fresh), Rutab (half-ripe), and Tamar (fully ripe or dried), depending on the cultivar. Cultivars are also classified into soft, semidry, and dry based on moisture content, texture, appearance, and sugar levels. There are two types of dried dates: Tamar, which can be naturally tree-dried or sun-dried, and Chuhhara (or Khalal matbookh), where Khalal-stage dates are boiled and then sun-dried. The latter method, commonly used in Pakistan, extends shelf life while preserving nutritional value (Khalid et al., 2017). Dietary fibers in dates may help combat hypercholesterolemia, a key risk factor for chronic heart disease. They reduce cholesterol absorption and reabsorption in the intestine and inhibit cholesterol production in the liver by producing short-chain fatty acids. Additionally, dietary fibers in dates can lower the atherogenic potential of LDL particles, which may consequently reduce the risk of cardiovascular disorders (Younas et al., 2020). Dates offer various therapeutic benefits and possess significant nutritional value, making them a natural remedy for a range of non-communicable diseases and disorders. Their compositional analysis reveals that dates have notable antioxidant properties, primarily due to their ability to scavenge free radicals. Key antioxidants present in both the flesh and seeds of dates include carotenoids, polyphenols, and tannins. Dates also exhibit strong potential against chronic inflammatory diseases and are used to cure anti-inflammatory disorders, including arthritis, asthma, chronic obstructive pulmonary diseases, and gastrointestinal disturbances (Yasin et al., 2015). Thanks to their anti-inflammatory properties, components of date fruits are utilized in the pharmaceutical industry for producing pain relief medications, including ingredients in paracetamol and ibuprofen (El-Far et al., 2019). Dates, as an anti-inflammatory agent, can help counteract nitric oxide (NO) and tumor necrosis factor-alpha (TNF-α), both of which are key contributors to inflammation. Methanolic and aqueous extracts from date flesh and seeds exhibit anti-inflammatory properties and can be utilized to alleviate swelling in patients with adjuvant arthritis. Dates are a staple in Egyptian cuisine, often eaten on their own, added to desserts, or used in savory dishes. They are particularly important during Ramadan, when they are traditionally eaten to break the fast.

10.2.4 *Jujube (*Ziziphus jujuba*)*

Approximately 40 papyri, papyrus fragments, and ostraca from ancient Egypt, dating between 1900 BC and 300 BC, have survived. These texts provide nearly 2,000 medical prescriptions, offering insights into how ancient Egyptians treated various illnesses (Pommerening, 2012). Since *Z. spina-christi* has not yet been thoroughly studied through pharmacological research, it presents a valuable opportunity for interdisciplinary exploration. This could begin by examining its use in ancient Egypt and extending to modern investigations, including in vitro and in silico analyses of its compounds. Jujube has a long history of use in ancient Egypt. It is believed that Egyptians recognized its medicinal properties, using it for its soothing and rejuvenating effects. The fruit, leaves, and bark of the *Ziziphus* tree have been traditionally used in various remedies, especially to treat digestive issues, skin ailments, and respiratory problems. In rural and desert areas of Egypt, jujube fruit is commonly eaten. It is appreciated for its sweet, slightly tangy flavor and chewy texture. The tree grows well in Egypt's arid and semi-arid regions, especially in areas like **Sinai** and **Upper Egypt**, where it thrives in poor soil conditions and intense heat. It's often found in small home gardens or wild groves. The ethnic uses of jujube fruit are given in Table 10.3.

Table 10.3 Ethnic uses of Jujube (Ziziphus jujuba) in Egypt

Category	Ethnic use	Details
Culinary uses	Raw consumption	Eaten fresh during harvest season, especially in rural and desert areas. Sweet, tangy flavor.
	Dried jujube	Dried for preservation, used as a snack or ingredient in traditional dishes
	Desserts and confections	Sometimes used in traditional sweets like halva, mixed with honey or nuts in some regions
	Herbal beverages	Boiled to create infusions or teas with calming, soothing effects, often consumed for relaxation or as a remedy for common ailments
Medicinal uses	Digestive aid	Used to treat indigestion, improve digestion, and soothe the stomach
	Respiratory remedy	Tea or decoction is used for sore throats, coughs, and cold symptoms
	Topical application	The fruit or leaves are applied topically for skin issues such as irritation or rashes, and to promote hair health
	General health tonic	Used in traditional medicine to boost immunity and general well-being, often as part of folk remedies
Religious and cultural uses	Symbol of protection (Sidr)	The tree is associated with religious significance in Islam, symbolizing paradise and protection
	Traditional knowledge and practices	Passed down in rural communities, especially in Bedouin cultures, where the tree's uses are tied to heritage and subsistence
Ecological and agricultural uses	Drought-resistant crop	Jujube is grown in arid and semi-arid regions of Egypt, particularly in the Sinai and Upper Egypt, due to its ability to thrive in harsh, dry conditions
	Sustainable agriculture	Increasing interest in its use in agroforestry and sustainable farming practices, as it requires minimal water and care

(*Continued*)

Table 10.3 (Continued) Ethnic uses of Jujube (Ziziphus jujuba) in Egypt

Category	Ethnic use	Details
Preservation techniques	Drying and storage	Traditionally dried to prolong its shelf life, making it available for consumption outside of the harvest season
Nutritional uses	High in nutrients	Rich in vitamin C, antioxidants, potassium, and magnesium, considered a health-boosting superfood in folk nutrition

10.3 Conclusion

In conclusion, Egypt's strategic geographic location and millennia-old history have profoundly shaped its rich and diverse cultural, agricultural, and culinary heritage. The country's fertile lands, sustained by the Nile River, have not only supported the development of one of the world's earliest civilizations but also fostered the growth of a complex and enduring food culture. Ancient Egyptian gastronomy, deeply tied to both survival and spiritual beliefs, placed significant emphasis on the consumption of plant-based foods, with historical evidence revealing a predominantly vegetarian diet among various age groups. The strong connection between food, health, and cultural identity persists in modern Egypt, where traditional dietary practices, influenced by Pharaonic, Coptic, Greco-Roman, and Islamic traditions, continue to play a pivotal role in the daily lives of its people. Medicinal plants and fruits, which have been integral to Egypt's ethnobotanical traditions for thousands of years, further illustrate the strong link between ancient practices and contemporary health care. The use of plants such as dates, grapes, jujube, and baobab for their therapeutic properties reflects a longstanding reliance on natural resources for both nutrition and medicine. As global interest in herbal medicine and functional foods continues to grow, Egypt's wealth of medicinal plants, with their demonstrated efficacy in treating a variety of ailments, offers significant potential for both local and international markets. However, the preservation and sustainable utilization of these plant species remain critical challenges, especially in the face of declining biodiversity and the loss of traditional knowledge. Moreover, the increasing recognition of traditional foods and medicinal fruits highlights the importance of safeguarding Egypt's culinary heritage as an essential aspect of its cultural identity. Efforts to document, study, and conserve these practices are crucial for ensuring that future generations can continue to benefit from the wisdom of the past. Egypt's enduring legacy in both cuisine and ethnomedicine not only enriches its national identity but also contributes to the global understanding of the interconnection between food, culture, and health. By bridging the gap between ancient practices and modern scientific research, Egypt can play a pivotal role in advancing the global appreciation of traditional medicinal fruits and their potential contributions to contemporary health and well-being.

References

Abdel-Azim, N. S., Shams, K. A., Shahat, A. A., El Missiry, M. M., Ismail, S. I., & Hammouda, F. M. (2011). Egyptian herbal drug industry: challenges and future prospects. *Research Journal of Medicinal Plants*, *5*(2), 136–144.

Amer, H. M., & Mohammad, A. A. (2022). Medicinal plants and their validation challenges in traditional Egyptian medicine. *Journal of Applied Pharmaceutical Science*, *12*(3), 023–033.

Chadare, F. J. (2010). *Baobab (Adansonia digitata L.) Foods from Benin: Composition, Processing and Quality*. Wageningen University and Research.

El-Far, A. H., Oyinloye, B. E., Sepehrimanesh, M., Allah, M. A. G., Abu-Reidah, I., Shaheen, H. M., … & Mousa, S. A. (2019). Date palm (Phoenix dactylifera): novel findings and future directions for food and drug discovery. *Current Drug Discovery Technologies*, *16*(1), 2–10.

El-Hamid, A., Mohamed, N. S., Makled, S. M. S., & Abd Elmonem, S. (2019). Determinants of production and export for some medicinal and aromatic plants in Egypt. *Arab Universities Journal of Agricultural Sciences*, *27*(2), 1351–1369.

Elsabawy, M. N. E. (2012). Medicinal and aromatic crops in Egypt: a study in medical geography. *ARPN Journal of Science and Technology*, *2*, 207–221.

Khalid, S., Khalid, N., Khan, R. S., Ahmed, H., & Ahmad, A. (2017). A review on chemistry and pharmacology of Ajwa date fruit and pit. *Trends in Food Science & Technology*, *63*, 60–69.

Laborde, S. (2017). *Following the Sun: A Practical Guide to Egyptian Religion*. Lulu Press, Inc.

Lisao, K., Geldenhuys, C. J., & Chirwa, P. W. (2017). Traditional uses and local perspectives on baobab (Adansonia digitata) population structure by selected ethnic groups in northern Namibia. *South African Journal of Botany*, *113*, 449–456.

Mahmoud, T., & Gairola, S. (2013). Traditional knowledge and use of medicinal plants in the Eastern Desert of Egypt: a case study from Wadi El-Gemal National Park. *Journal of Medicinal Plants*, *1*(6), 10–17.

Mehdawy, M., & Hussein, A. (2010). *The Pharaoh's Kitchen: Recipes from Ancient Egypt's Enduring Food Traditions*. American Univ in Cairo Press.

Parsa, H. G. (1998). Cultural heritage of American food habits and implications for the hospitality industry. *Marriage & Family Review*, *28*(1–2), 23–48.

Pommerening, T. (2012). Altägyptische Rezepte–Eine diachrone Betrachtung. *Geschichte der Pharmazie*, *64*(3), 33–38.

Ramli, A. M., Zahari, M. S. M., Halim, N. A., & Aris, M. H. M. (2016). The knowledge of food heritage Identithy in Klang Valley, Malaysia. *Procedia-Social and Behavioral Sciences*, *222*, 518–527.

Shemesh, A. O. (2020). Food, memory and cultural-religious identity in the story of the 'desirers' (Nm 11: 4–6). *HTS Teologiese Studies/Theological Studies*, *76*(3). doi:10.4102/hts.v76i3.6158

Yasin, B. R., El-Fawal, H. A., & Mousa, S. A. (2015). Date (Phoenix dactylifera) polyphenolics and other bioactive compounds: a traditional Islamic remedy's potential in prevention of cell damage, cancer therapeutics and beyond. *International Journal of Molecular Sciences*, *16*(12), 30075–30090.

Younas, A., Naqvi, S. A., Khan, M. R., Shabbir, M. A., Jatoi, M. A., Anwar, F., … & Aadil, R. M. (2020). Functional food and nutra-pharmaceutical perspectives of date (Phoenix dactylifera L.) fruit. *Journal of Food Biochemistry*, *44*(9), e13332.

Zahra'u, B., Mohammed, A. S., Ghazali, H. M., & Karim, R. (2014). Baobab tree (Adansonia digitata L) parts: nutrition, applications in food and uses in ethno-medicine–a review. *Annals of Nutritional Disorders & Therapy*, *1*(3), 1011.

Index

Note: **Bold** page numbers refer to tables and *italic* page numbers refer to figures.

acaricidal activity 66
Actinidia deliciosa 119
Adansonia digitata 184–185
agricultural marketing resource centre (AMRC) 123
antibacterial activity 63
antibacterial property 95–96
anti-cancer 76
anti-colitis activity 76
anticonvulsant effect 83–84
antifungal activity 63
anti-fungal property 98
antihirustism activity 66
antihyperglycemic activity 77
anti-inflammatory activity 65
anti-inflammatory effect 85
anti-ischemic property 97
antimicrobial activity 77
anti-migratory property 97
anti-obesity activity 76
antioxidant activity 63–64
antioxidant property 94–95
antioxidants extraction
 caraway (*Carum carvi*) 77–79
 Foeniculum vulgare 69–72
 Pimpinella anisum 90–91
 Rhus coriaria 99–100
anti-oxidative stress activity 77
antitubercular activity 75–76
antitumor activity 65–66
appetite-suppressing properties 74–75
apricot 12
 drying 10–11, *11*
 squash 9–10, **10**, *10*
aqueous extraction 19, *24*
 advantage and application 24
ascorbic acid 3, 72, 128, 129, 140, 157, 159, 162
Assyrian plums *see Cordia myxa*
avatiḥim 181

Bahrain
 medicinal fruits 135, 137
  challenges 142
  Citrullus colocynthis 137–138, **138**
  Datura fastuosa 141–142
  nutraceuticals *see* nutraceuticals
  Ochradenus baccatus Delile 139–140
  Prosopis farcta 138–139
  Zizphus spina-christi 140
Baobob (Adansonia digitata) 184–185
barberry (*Berberis vulgaris*) 14–15, 26, 47
Berberis integerrima 15
Berberis vulgaris see barberry (*Berberis vulgaris*)
betzalim 181
bitter apple fruit 11
black mulberry (*Morus nigra*) 14, 15, 18
blood pressure lowering 75
butylated hydroxyanisole (BHA) 64, 70, 71, 85, 86, 90
butylated hydroxytoluene (BHT) 64, 70, 71, 85, 86, 90, 99

caraway (*Carum carvi*) 72–73, *74*
 antioxidants extraction 77–79
 nutritional uses 73, **73**
  anti-cancer 76
  anti-colitis activity 76
  antihyperglycemic and hypolipidemic activity 77
  antimicrobial activity 77
  anti-obesity activity 76

caraway (*Carum carvi*) (*cont.*)
anti-oxidative stress activity 77
antitubercular activity 75–76
appetite-suppressing properties 74–75
blood pressure lowering, vasodilator and cardiac-modulatory 75
cardiovascular disease 76
chemical compounds 74
hypolipidemic effect 77
hypothyroidism 75
irritable bowel syndrome 75
oral mucositis 75
potential side effects 77
cardiac-modulatory 75
cardiovascular disease 18, 61, 76, 98, 133, 154, 166
Carica papaya 7
chemical compounds 16, 74, 140
cherry (*Prunus avium*) 18–19
Citrullus colocynthis 137–138, **138**
bioactive properties 11–12
limitations 12
cold pressing 101
commercial dairy dessert (CDD) 151
Cordia myxa 1–3, *3*
bio-active compounds
ascorbic acid 3
γ-Sitosterol 3
linoleic acid 3
stigmasterol 2
cornelian cherry (*Cornus mas L.*) 18
Cornus mas L. see cornelian cherry (*Cornus mas L.*)
Crataegus elbursensis 16
Crocus sativus 14, 19
culinary uses 58, 158, 180, 181, 189
Cynomoriaceae 4
Cynomorium coccineum 4
cytoprotection, *Foeniculum vulgare* 65–66

daga (fish) 181
date fruit fiber concentrates (DFC) 153
date palms (*Phoenix dactylifera*) 16–17, 26, 28, 124–127, 146, 149, 152, 153, 167, 186–187
antioxidant and anuimicrobial properties 127
nutritional characteristics 125, 127
date powder (DP) 151
date syrup (DS) 150–153
Datura fastuosa 141–142
diabetes patients 56
Pimpinella anisum 87
DNA protective effect, *Rhus coriaria* 97–98
Dombeya arabica 3
dymenorrhea, *Pimpinella anisum* 85

Egypt 3, 4, 79
medicinal fruits 180–182, **183–184**, 189–190
Baobob (Adansonia digitata) 184–185
Dates (Phoenix dactylifera) 186–187
jujube (*Ziziphus jujuba*) 187, **188–189**
Wine grapes (Vitis vinifera) 185–186
Elaeagnus angustifolia L. 17
enzyme-assisted extraction (EAE) 15, 31, *32*, 33, 102
advantage and application 31, 33
essential oils (EOs) 16, 59–66, 69–72, 74–80, 82–86, 90, 99–102, 129
estragole, *Foeniculum vulgare* 60, 69, *70*, 72, 80, 81, 90

Ficus carica see figs (*Ficus carica*)
Ficus sycomorus 4
figs (*Ficus carica*) 16, 117, *118*, 124, 128, 177–179, *178*
antioxidant and antimicrobial properties 128
mulberry 4
nutritional characteristics 128
in vitro gastrointestinal model system 177–179
Foeniculum vulgare 57–58
acaricidal activity 66
antibacterial activity 63
antifungal activity 63
antihirustism activity 66
anti-inflammatory activity 65
antioxidant activity 63–64
antioxidants extraction 69–72
culinary uses 58
cytoprotection and antitumor activity 65–66
essential oil 59–62
estragole 60, 69, *70*, 72, 80, 81, 90
hepatoprotective activity 65
human liver cytochrome P450 3A4 66–67, *67*
molecular structures 59–61, *59–62*
oestrogenic activity 65
phytochemistry 59
potential mechanism 67, **68**

spice safety 69
use in local and traditional medicine 58–59
uterine contraction 66
fruit global value chain 122, *123*
importance 123

γ-Sitosterol 3
gas chromatographic-mass spectral (GC–MS) 71, 90, 92, 161
gas chromatography (GC) 69, 71, 80, 90, 92, 93, 99, 161
gastrointestinal system 84
Glossonema edule 4
Glossostemon bruguieri 3
grapes (*Vitis vinifera*) 18, 33, 163, 185, 186, 189
grenadine 158–159, 168

ḥatzir 181
hawthorn (*Crataegus elbursensis*) 16
HDL 177
hepatoprotective activity, *Foeniculum vulgare* 65
High Pressure Processing (HPP) 158
high storage temperatures (HSJ) 158
hot solvent extraction 100, 101
human liver cytochrome P450 3A4 66–67, *67*
hydrostatic pressure (HHP) 155, 161
hypoglycemic property, *Rhus coriaria* 96–97
hypolipidemic activity 77
hypothyroidism 72, 75

insecticidal effect, *Pimpinella anisum* 86–87
Iran
colour pigment 33, *35*
aqueous extraction 19, 24, *24*
enzyme-assisted extraction (EAE) 31, *32*, 33
and health benefits 19, **20–23**
microwave-assisted extraction (MAE) 33, *34*
reflux extraction 26, *27*
solvent extraction *25*, 25–26
supercritical fluid extraction (SFE) 28, *29*
ultrasound-assisted extraction (UAE) 28, *30*, 31
medicinal fruits 14–15
barberry (*Barberis spp.*) 15
cherry (*Prunus avium*) 18–19
cornelian cherry (*Cornus mas L.*) 18
date palms (*Phoenix dactylifera*) 16–17
figs (*Ficus carica*) 16
grapes (*Vitis vinifera spp.*) 18
hawthorn (*Crataegus elbursensis*) 16
Iranian Qara Qara (*Ribes rubrum*) 17
jujube (*Ziziphus jujuba*) 19
mulberry (*Morus spp.*) 18
myrtle berries (*Myrtus communis L.*) 15–16
pistachio (*Pistacia atlantica*) 16
pomegranates (*Punica granatum*) 17
saffron (*Crocus sativus*) 19
silverberry (*Elaeagnus angustifolia L.*) 17
natural colourants extraction 33, 36, **37–46**
Iranian Qara Qara (*Ribes rubrum*) 17
Iraq
medicinal fruits 56–57
caraway (*Carum carvi*) *see* caraway (*Carum carvi*)
Foeniculum vulgare see Foeniculum vulgare
Pimpinella anisum see Pimpinella anisum
Rhus coriaria see Rhus coriaria
irritable bowel syndrome, caraway (*Carum carvi*) 75

Jordan
medicinal fruits 176
fig fruit 177–179, *178*
pomegranate fruit 176–177, **178**
jujube (*Ziziphus jujuba*) 19, 118, *119*, 125, 128–129, 187, **188–189**
antioxidant and antimicrobial properties 129
nutritional characteristics 128–129

kishuim 181
kiwi 116, 119
Kuwait
medicinal fruits 146
anardana and pomegranate syrup/grenadine 158–159
Phoenix dactylifera (*dates*) 149–153
pomegranate (*Punica granatum*) *see* pomegranate (*Punica granatum*)
Ziziphus spina-christi (*sidr*) 146–149

LDL 164, 177, 187
Lebanon
medicinal fruits
fig fruit 117, *118*
jujuba 118, *119*
kiwi 119
pomegranate fruit 116–117, **117**
linoleic acid 3, 64, 71, 86, 90, 91, 94
low storage temperatures (LSJ) 158

medicinal fruits
Bahrain 135, 137
challenges 142
Citrullus colocynthis 137–138, **138**
Datura fastuosa 141–142
nutraceuticals *see* nutraceuticals
Ochradenus baccatus Delile 139–140
Prosopis farcta 138–139
Zizphus spina-christi 140
Egypt 180–182, **183–184**, 189–190
Baobob (Adansonia digitata) 184–185
Dates (Phoenix dactylifera) 186–187
jujube (*Ziziphus jujuba*) 187, **188–189**
Wine grapes (Vitis vinifera) 185–186
Iran *see* Iran, medicinal fruits
Iraq 56–57
caraway (*Carum carvi*) *see* caraway (*Carum carvi*)
Foeniculum vulgare see Foeniculum vulgare
Pimpinella anisum see Pimpinella anisum
Rhus coriaria see Rhus coriaria
Jordan 176
fig fruit 177–179, *178*
pomegranate fruit 176–177, **178**
Kuwait 146
anardana and pomegranate syrup/ grenadine 158–159
Phoenix dactylifera (*dates*) 149–153
pomegranate (*Punica granatum*) *see* pomegranate (*Punica granatum*)
Ziziphus spina-christi (*sidr*) 146–149
Lebanon
fig fruit 117, *118*
jujuba 118, *119*
kiwi 119
pomegranate fruit 116–117, **117**
Muscat 122
agricultural marketing resource centre (AMRC) 123
dates (Phoenix dactylifera) 124–127
fig (Ficus carica) 124, 128
fruit and vegetables global value chain 122, *123*
jujube (*Ziziphus jujuba*) 125, 128–129
Oman crops 124–125
pomegranate (Punica granatum) 124, 127–128
USDA 124
Saudi Arabia *see* Saudi Arabia
United Arab Emirates (UAE) *see* United Arab Emirates (UAE)
menopausal hot flashes 90
Pimpinella anisum 85
microwave-assisted extraction (MAE) 33, *34*, 57, 100, 101
advantage and application 33
mixture dairy dessert (MDD) 151
morphine dependence 90
Pimpinella anisum 85
Morus nigra 14, 15, 18
mulberry (*Morus spp.*) 18
Morus nigra 14–15, 18
Muscat
medicinal fruits 122
agricultural marketing resource centre (AMRC) 123
dates (Phoenix dactylifera) 124–127
fig (Ficus carica) 124, 128
fruit and vegetables global value chain 122, *123*
jujube (*Ziziphus jujuba*) 125, 128–129
Oman crops 124–125
pomegranate (Punica granatum) 124, 127–128
USDA 124
muscle relaxant effect, *Pimpinella anisum* 83
myrtle berries (*Myrtus communis L.*) 15–16
Myrtus communis L. see myrtle berries (*Myrtus communis L.*)

natural colourants extraction 33, 36, **37–46**
non-mutagenic property, *Rhus coriaria* 96
non-thermal extraction methods
Rhus coriaria 100–101
techniques 101–102
non-timber forest products (NTFPs) 184
nutraceuticals 4, 132, 133, 137, 142, 143, 146, 148, 166, 167
benefits 133, *135*
biological activities 125, **126**, 135, **136**
classification 132, *133*
composition and color 135, **135**

fruits 134, 135
types 132, **134**

Ochradenus baccatus Delile 139–140
oestrogenic activity, Foeniculum vulgare 65
Oman crops 124–125
ORAC 152, 153, 157
oral mucositis, caraway (Carum carvi) 75

P450 3A4, Foeniculum vulgare 66–67, 67
papaya fruit
jam 7–8, **8**, *8*
juice concentrate 8–9, *9*
PGE2 66
Phoenix dactylifera (dates) 16–17, 124–127, 146, 149–153, 186–187
Pimpinella anisum 79, 100–102
antioxidants extraction 90–91
chemical constituents 80–81
pharmacological properties 81–82, 87, **87–89**
analgesic and anti-inflammatory effect 85
anticonvulsant effect 83–84
antioxidant activity 85–86
diabetes patients 87
dymenorrhea 85
gastrointestinal system 84
insecticidal effect 86–87
menopausal hot flashes 85
morphine dependence 85
muscle relaxant effect 83
traditional medicine 80
pistachio (*Pistacia atlantica*) 16
pomegranate (*Punica granatum*) 14, 17, 26, 31, 33, 36, 47, 57, 116–117, **117**, 124, 127–128, 146, 153–155, 176–177, **178**
antioxidant and antimicrobial properties 127–128
beverages 160–161
concentrate 159–160
confectionery 162–163
juice and jam 155–158
molasses 159
nutritional characteristics 127
powder and rind powder 164–166
seed oil 166
seed powder 167
waste utilization 167
wine 163–164
pomegranate fruit concentrate (PFC) 160
pomegranate juice concentrate (PJC) 160
pomegranate molasses (PM) 159
pomegranate peel powder meal (PPPM) 165
pomegranate rind powder extract (PRPE) 166
pomegranate seed powder (PSP) 167
Prosopis farcta 137–139
Prunus armeniaca 9, 12
Prunus avium 18–19
Punica granatum see pomegranate (*Punica granatum*)

Red Thumb 4
reflux extraction 26, *27*
advantage and application 26
Rhus coriaria 91, 102
antibacterial property 95–96
anti-fungal property 98
anti-ischemic property 97
anti-migratory property 97
antioxidant property 94–95
antioxidants extraction 99–100
DNA protective effect 97–98
hypoglycemic property 96–97
non-mutagenic property 96
phytochemistry 91–93, *92*
thermal and non-thermal extraction methods 100–101
traditional medicine use 93–94
Ribes rubrum 17
Russian olive 17

saffron (*Crocus sativus*) 14, 19, 28, 33, 85
Saudi Arabia 139, 154
medicinal fruits 1, **2**
Cordia myxa 1–3, *3*
Cynomorium coccineum 4
Ficus sycomorus 4
Glossonema edule 4
Glossostemon bruguieri 3
Secretariat of the Convention on Biological Diversity 182
shumim 181
Sicilian sumac *see Rhus coriaria*
sidr fruit pulp (SFP) 147
silverberry (*Elaeagnus angustifolia L.*) 17
solvent extraction 15, *25*, 25–26, 57, 100, 101
advantage and application 26
steam distillation 100–101
stigmasterol 2

supercritical fluid extraction (SFE) 15, 28, *29*, 57, 74, 101
  advantage and application 28
sycamore fig 4, 185
syrup dairy dessert (SDD) 151

thermal extraction methods, *Rhus coriaria* 100–101
TNBC cells 76
total anthocyanin content (TAC) 156, 158
total phenol content (TPC) 141, 147, 149, 151, 152, 156, 161–164
total soluble solids (TSS) 8, 157, 159, 162

ultrasound-assisted extraction 15, 28, *30*, 31, 101, 154
  advantage and application 28, 31
United Arab Emirates (UAE)
  medicinal fruits
    apricot *see* apricot
    citrullus colocynthis *see* citrullus colocynthis
    papaya fruit *see* papaya fruit
USDA 124
uterine contraction 66

vasodilator 100
  caraway (*Carum carvi*) 75
vegetables global value chain 122, 123, *123*
Vitis vinifera see grapes (*Vitis vinifera*)

wine grapes (*Vitis vinifera*) 185–186
World Health Organization 122

Ziziphus jujuba (jujube) 19, 118, *119*, 125, 128–129, 187, **188–189**
Ziziphus spina-christi (*sidr*) 140, 146–149, 187

For Product Safety Concerns and Information please contact our EU representative GPSR@taylorandfrancis.com
Taylor & Francis Verlag GmbH, Kaufingerstraße 24, 80331 München, Germany

www.ingramcontent.com/pod-product-compliance
Lightning Source LLC
LaVergne TN
LVHW020638300926
849161LV00004B/747
* 9 7 8 1 0 3 2 5 8 9 4 2 8 *